Home Automation
with Raspberry Pi®

About the Author

DONALD NORRIS has a degree in electrical engineering and an MBA with a specialization in production management. He is currently teaching both undergraduate and graduate IT courses at Southern New Hampshire University. He has also created and taught several robotics courses there, in addition to many different computer science and engineering courses. He has over 35 years of teaching experience as an adjunct professor at a variety of colleges and universities.

Mr. Norris retired from civilian government service with the U.S. Navy, where he specialized in acoustics related to nuclear submarines and associated advanced digital signal processing. Since then, he has spent more than 20 years as a professional software developer using C, C#, C++, Python, and Java, as well as 5 years as a certified IT security consultant.

Mr. Norris started a consultancy, Norris Embedded Software Solutions (dba NESS LLC; www.nessllc.net), that specializes in developing application solutions using microprocessors and microcontrollers. He likes to think of himself as a perpetual hobbyist and geek, and is always trying out new approaches and out-of-the-box experiments. He is a licensed private pilot, photography buff, amateur radio operator, and avid health enthusiast.

Besides the present book, Mr. Norris is the author of seven other McGraw-Hill TAB books, dealing with subjects ranging from learning programming to Raspberry Pi projects to building your own quadcopter.

Home Automation with Raspberry Pi®

Projects Using Google Home™, Amazon Echo®, and Other Intelligent Personal Assistants

Donald Norris

New York   Chicago   San Francisco   Athens   London   Madrid
Mexico City   Milan   New Delhi   Singapore   Sydney   Toronto

Library of Congress Control Number: 2019931445

This book is printed on acid-free paper.

Sponsoring Editor Lara Zoble	**Copy Editor** James Madru
Editorial Supervisor Stephen M. Smith	**Proofreader** Alison Shurtz
Production Supervisor Pamela A. Pelton	**Indexer** Jack Lewis
Acquisitions Coordinator Elizabeth M. Houde	**Art Director, Cover** Jeff Weeks
Project Manager Patricia Wallenburg, TypeWriting	**Composition** TypeWriting

This is book is dedicated to my parents, Harry and Esther Norris, who provided me with a loving and nurturing childhood. Their guidance and wisdom greatly influenced how I matured and helped me to develop a healthy scientific curiosity. I also wish to acknowledge my brothers, Herb and Ken, with whom I shared my childhood and who helped me develop a sense of self-confidence and an ability to be my own person. Finally, I wish to acknowledge my sister, Lori, who was born when I was about ready to leave my family and strike out on my own. Lori has consistently demonstrated to me the ability to achieve remarkable goals, culminating with her attaining the role of adult nurse practitioner (ANP).

Contents

viii   Contents

Preface

THIS BOOK IS ALL ABOUT HOW YOU, as a maker, can help automate your home or business to both improve the quality of your life and coincidentally achieve some efficiencies. The latter may be actual energy savings or may be simply improving the everyday flow of personal activities. I also included using the Raspberry Pi as a principal controller in most of this book's projects because it so inexpensive yet provides amazing capabilities and functionalities when implementing home automation (HA) solutions.

I have written about HA projects in several of my earlier maker books, which also included the Raspberry Pi as the main controller. However, in this book, I have included some rather extensive discussions regarding personal voice assistants and their role in HA projects. This type of device is fairly new to the marketplace and includes the rather well-known Amazon Echo and Google Home devices. The inclusion of these devices in this book was predicated on my realization that they are rapidly becoming the consumer's favorite choice when interacting with HA systems. The days of turning a thermostat dial or flipping a light switch are rapidly declining given the ubiquitous nature of these new voice-activation devices.

The first few chapters of this book are devoted to exploring how various voice-activation devices function with the Raspberry Pi. There are significant differences between how Amazon Echo devices interact with the Raspberry Pi and how Google Home devices interact. I try to explain how you can successfully interface with both device types, while at the same time I point out the pros and cons of the two approaches. I am confident that if you successfully follow my multiple demonstrations, you will become quite adept at interfacing with either device class.

Chapter 5, which follows all the material on voice-activation devices, concerns HA operating systems, which are an extremely useful adjunct in the implementation of a workable HA system. I provide a good survey of all the major, currently available HA operating systems (OSs), but such surveys are highly volatile, and there will likely be significant changes by the time this book is published. Nonetheless, the fundamentals of a good HA OS are unchanging, and, hopefully, my discussion in this area will provide you with some good guidance regarding the selection of an appropriate OS that meets your requirements. Chapter 5 also includes a working demonstration of one of the most popular HA OSs, which should help you really understand how this software functions and let you make a good decision on whether or not to pursue this option as an HA solution.

Chapter 6 describes a Z-Wave-enabled HA system. Z-Wave is a very popular way to create communications links among separate or distributed HA components. There are quite a few Z-Wave-compliant device manufacturers in business offering many different and varied

Z-Wave-enabled components. I feel that it is important for you to be aware of this particular communications protocol because of its impressive popularity and many readily available devices and components. I also provide a chapter demonstration in which I interface a Raspberry Pi directly into a Z-Wave system. This allows you to create your own custom control scripts for many different Z-Wave devices.

An open-source alternative to the Amazon Echo and Google Home devices is the topic of Chapter 7. The open-source device I discuss is named *Picroft*, and it is a Raspberry Pi–hosted variant on a parent device named *Mycroft*. Basically, Mycroft and Picroft are synonymous: Mycroft is an actual open-source device, which can be purchased, and Picroft is the software image loaded onto a Raspberry Pi. In reality, the Mycroft device contains a Raspberry Pi running precisely the same software as that contained in the Picroft image. Given this fact, I constantly interchange the names Mycroft and Picroft in the chapter without a loss of understanding or context. My purpose in discussing Mycroft is to present you with a lower-cost option to the Amazon and Google devices. However, the cost difference is really quite marginal when considering that you will need both an external USB microphone and speaker/amplifier to make up a working Picroft system. If you add all these extra costs to the cost of a Raspberry Pi, you will likely pay about the same as you would if you purchased an Amazon Echo Dot or a Google Home Mini. However, if the maker in you is up to the task, it is always fun to build your very own open-source voice assistant.

A trip into the artificial intelligence (AI) realm is the subject of Chapter 8. There I cover the principles that govern how AI fuzzy logic (FL) may be applied to a home heating, ventilation, and air-conditioning (HVAC) system. I discuss in great detail how to use a multistep procedure to design a workable FL system. The good news is that no additional or expensive components are required for a Raspberry Pi–powered HVAC FL controller. The only requirement is to create and load the FL code into the Raspberry Pi. There is an actual FL demonstration that simulates how a real HVAC system would function. I use light-emitting diodes (LEDs) to indicate when appropriate heating and/or cooling commands are generated.

Chapter 9 is a kind of a catch-all where I cover how to use a variety of sensors seen commonly in several different HA systems, including HVAC and security-type systems. This chapter provides reasonable insights into what to consider when selecting a sensor and how to design appropriate interfaces between the sensors and the controller (which, of course, in this chapter is a Raspberry Pi).

Chapter 10 describes an HA security system that employs a remotely located sensor from the Raspberry Pi main controller. I use a very nice wireless data link system named *XBee*, which provides the capability of sending not only a binary on/off signal but also actual sensor data values, if needed. The XBee subsystem is controlled by an Arduino Uno microcontroller, which provides me with an opportunity to introduce the coprocessor concept into our discussion of HA system design. Sometimes a Raspberry Pi cannot "do it all" and needs some assistance. Timing is a very important feature for any communications link. However, the Linux OS running on a Raspberry Pi is *asynchronous*, meaning that there is no guarantee that the computer will be available to process incoming communications data. Meanwhile, the Uno does provide the immediate and continuous attention required by the communications link.

Chapter 11 is a brief one in which I discuss some important ideas on how to integrate separate HA systems so that the user has a "unified" view of an overall HA system and how to provide one-stop control. I also discuss how scripts or macros executed on an HA controller can significantly improve the overall HA user experience.

Donald Norris

## CHAPTER 1

Designing and Building Home Automation Projects

I LIKE TO THINK OF THIS CHAPTER as providing the prerequisite knowledge to allow you to build home automation (HA) projects as detailed in this book and other sources. It also seems that learning about the "big picture" is always a useful approach before digging into specifics and fine-grain details. The next section details a generalized approach to designing and implementing an HA solution.

I will include a Parts List at the beginning of each chapter so that readers can identify what parts will be required to duplicate the projects and/or demonstrations presented in that chapter. The Parts Lists also provide suggested sources that are current at the time of this writing. Actual parts that are needed will depend on what each reader already has in his or her parts bin. This first list also includes parts and components that are common to other Parts Lists, such as power supplies and HDMI cables. I will not include these common items in later Parts Lists because they are self-evident.

Parts List

Item	Model	Quantity	Source
Raspberry Pi 3	Model B	1	mcmelectronics.com adafruit.com digikey.com mouser.com farnell.com
T-Cobbler	40-pin P/N 2029	1	adafruit.com
Solderless breadboard with 830 tie points	Commodity	1	mcmelectronics.com adafruit.com digikey.com mouser.com farnell.com
5-volt (V), 2.5-ampere (A) power supply with micro USB connector	Commodity	1	mcmelectronics.com adafruit.com digikey.com mouser.com farnell.com
HDMI-to-HDMI cable, 1 meter (m)	Commodity	1	amazon.com

(continued on next page)

Item	Model	Quantity	Source
LEDs, various colors	Commodity	Varies	mcmelectronics.com adafruit.com digikey.com mouser.com farnell.com
330-ohm (Ω), ¼-watt (W) resistor	Commodity	Varies	mcmelectronics.com adafruit.com digikey.com mouser.com farnell.com
2N3904 NPN transistor	Commodity	Varies	mcmelectronics.com adafruit.com digikey.com mouser.com farnell.com
Jumper wire package	Commodity	1	adafruit.com
USB keyboard	Amazon Basic	1	amazon.com
USB mouse	Amazon Basic	1	amazon.com
HDMI monitor	Commodity	1	amazon.com
Google AIY Voice Kit for Raspberry Pi	3602	1	adafruit.com

Generalized HA Design Approach

The first step in any HA project, except for the most trivial one, is to define the requirements. This means that definite and detailed project requirements should be written in a clear and nonambiguous manner. For example, simply stating "to turn on the lights" would be incorrect because there is no mention of what specific lights are to be turned on or activated, how they are to be activated, or how long they should stay on. A better version for the requirements statement might be "to turn on the front porch lights for one minute by voice activation." In this case, a specific light is identified as well as an operational time and an activation mode. Each of the phrases in the requirement statement will naturally lead to specific HA implementations. Creating definitive, clear, and precise requirements should help you to develop appropriate HA solutions to meet your individual needs in an HA system.

The architecture of an HA solution depends very much on the device types to be controlled and their locations within the home. For instance, controlling a thermostat is a very different process from activating a light. Different control technologies are involved in each case, which employ different means of interfacing with a particular technology. I will also be using a Raspberry Pi as the standard microcontroller for all the HA projects in this book, which will help to minimize any confusion regarding interfacing or data-communication issues. However, I will be using several different control technologies because some HA devices are widely separated, which typically requires a wireless control technology, whereas other devices are close to the controller, which might best be served by using a direct-wired approach. In a few select cases, I will use wired control lines for dispersed devices but only use two wires for the control data signals.

One very important design feature that you always must keep in mind when designing an HA system is *safety*. Any controlled device that uses a mains supply should always be controlled with a certified device that is appropriate to the country where it is being used. In the United States, this means that the mains control device should have Underwriters Laboratories (UL) approval. There are similar rating agencies in other countries where HA projects are designed and implemented. Using certified control devices will raise the cost of a project, but there is truly no other option when it comes to ensuring the safety and well-being of you and your family. None of this book's projects involve controlling mains-connected devices other than with UL-approved products.

Another important HA design feature that must be addressed is how the user will interact with the system. I mentioned voice activation in the earlier example because this approach seems to be the most popular at present. HA systems implemented just a few years ago relied on keypads and relatively simple liquid-crystal display (LCD)/ light emitting diode (LED) electronics to signal user interaction. Of course, most true home-brew HA systems used and still do use a traditional workstation coupled with a monitor, keyboard, and mouse. I will be using both the workstation and voice-recognition approaches in this book. Using a workstation is very advantageous when it comes to designing and developing specific HA solutions. The voice-recognition unit, or assistant, can then be integrated into the system once the initial design has been proven to work as expected. I will explain how the Google Voice Assistant functions a bit later in this chapter, but first I need to explain how to set up a Raspberry Pi such that it can function as an HA system controller.

Raspberry Pi Setup

You will need to set up a Raspberry Pi (RasPi) in order to duplicate this book's projects. I will show you how to set up a RasPi 3 Model B as a workstation that will host the applications required to implement a variety of HA solutions. Figure 1-1 shows the RasPi 3 Model B used in this book.

I should mention that a Raspberry Pi 3 Model B+ was just introduced by the Raspberry Foundation at the time of this writing. It is essentially the same as the Model B except for a slight speed improvement and some improvements in the wireless functions, none of which will have any impact on this book's projects. You can use either the B or B+ models without any software or hardware modifications.

I will not go into much detail about what makes up a RasPi single-board computer because that has already been adequately covered by many readily available books. I refer you to two of my earlier books, *Raspberry Pi Projects for the Evil Genius* and *Raspberry Pi Electronics Projects for the Evil Genius*, where I discuss in detail the architecture and makeup of the RasPi series of single-board computers. In this book, I use a RasPi 3 Model B in a workstation configuration, which is simply having the RasPi connected with a USB

Figure 1-1   Raspberry Pi 3 Model B.

keyboard, USB mouse, and HDMI monitor. The RasPi is powered by a 2.5-A, 5-V supply with a micro USB connector, as indicated in the Parts List. Now I will discuss how to set up secondary storage for a RasPi because this is critical to its operation.

A RasPi does not require a disk drive for storing and retrieving software applications and utilities, which includes an operating system (OS). The recent designs, within the last few years, all rely on using a pluggable micro SD card to serve this secondary storage function. It is also possible to connect a traditional disk drive to a RasPi, but it will only serve as an auxiliary storage device and not as the primary, persistent storage for the OS and boot partition. Next, I will show you how to download and install an OS on a micro SD card such that your RasPi can be booted to serve as a functional HA microcontroller.

The simplest way to obtain a programmed micro SD card is to purchase one from one of the RasPi suppliers listed in the Parts List. These cards are ready to go and only need to be configured to match your particular workstation configuration, which includes your private WiFi network. I will discuss the configuration process in a later section, but first I want to show you how to create your own micro SD card in case you do not have the means or desire to buy a preprogrammed card.

The software to be loaded is known as an *image* and is freely available from several online websites, with the recommended one being the Raspberry Pi Foundation site at raspberrypi.org. You will need to download the latest image from the Downloads section of the website. Two versions of the disk image are available. The first version is named *NOOBS*, which is short for "New Out Of the Box Software." The current NOOBS version available at the time of this writing is v2.7. This image, in reality,

is a collection of files and subdirectories that can be downloaded either using the BitTorrent application or simply as a raw Zip file. The BitTorrent and Zip downloads are approximately 1.2 gigabytes (GB) in size. The extracted image is 1.36 GB in size, but the final installed size is over 4 GB. This means that you will need to use at least an 8-GB micro SD card to hold the final image. However, I strongly recommend that you use at least a Class 10 card to maximize the data throughput with the operating RasPi.

The second image version available is named *Raspbian*, which is a Debian Linux distribution especially created for the RasPi. The currently available image is named *Stretch*. This version is also updated quite frequently, but I will show you how to ensure that you have the most up-to-date version in the configuration section. The Raspbian version may be downloaded using BitTorrent or as a Zip file with final image sizes similar to the NOOBS image.

A micro SD card must be loaded with the desired image after that image is downloaded. There are two ways to accomplish this task depending on the version downloaded. I discuss these ways in the following sections.

Writing the NOOBS Image to a Micro SD Card

The easiest way to create a bootable micro SD card is to use the downloaded and extracted NOOBS image. In this case, you will first need to format the micro SD card using an appropriate application compatible with your host computer. For Windows and Mac machines, use the following link to get the formatter program: www.sdcard.org/downloads/formatter_4/.

I used the Mac version without any problems. However, it is imperative that your micro SD card is properly formatted or else the NOOBS installation will not work.

2

Name	Date Modified	Size	Kind
os	March 31, 2018 at 9:19 AM	1.32 GB	Folder
recovery.rfs	March 14, 2018 at 8:38 AM	28.6 MB	Document
recovery7.img	March 14, 2018 at 8:38 AM	2.9 MB	NDIF Disk Image
recovery.img	March 14, 2018 at 8:38 AM	2.9 MB	NDIF Disk Image
recovery.elf	March 14, 2018 at 8:38 AM	672 KB	Document
overlays	March 14, 2018 at 8:37 AM	274 KB	Folder
bootcode.bin	March 14, 2018 at 8:38 AM	52 KB	MacBin...archive
defaults	March 14, 2018 at 8:38 AM	40 KB	Folder
bcm2710-rpi-3-b-plus.dtb	March 14, 2018 at 8:38 AM	18 KB	Document
bcm2710-rpi-3-b.dtb	March 14, 2018 at 8:38 AM	18 KB	Document
bcm2709-rpi-2-b.dtb	March 14, 2018 at 8:38 AM	17 KB	Document
bcm2710-rpi-cm3.dtb	March 14, 2018 at 8:38 AM	17 KB	Document
bcm2708-rpi-0-w.dtb	March 14, 2018 at 8:38 AM	16 KB	Document
bcm2708-rpi-b-plus.dtb	March 14, 2018 at 8:38 AM	16 KB	Document
bcm2708-rpi-b.dtb	March 14, 2018 at 8:38 AM	16 KB	Document
bcm2708-rpi-cm.dtb	March 14, 2018 at 8:38 AM	15 KB	Document
riscos-boot.bin	March 14, 2018 at 8:38 AM	10 KB	MacBin...archive
INSTRUCTIONS-README.txt	March 14, 2018 at 8:38 AM	2 KB	Plain Text
BUILD-DATA	March 14, 2018 at 8:38 AM	302 bytes	TextEdi...cument
recovery.cmdline	March 14, 2018 at 8:38 AM	99 bytes	Document
RECOVERY_FILES_DO_NOT_EDIT	March 14, 2018 at 8:38 AM	Zero bytes	TextEdi...cument

Figure 1-2   Contents of NOOBS_v2_7_0 folder.

The downloaded NOOBS file is named *NOOBS_v2_7_0.zip*, and when it is fully extracted, it is in a folder named *NOOBS_v2_7_0*. You must go into the folder and copy all the files and subdirectories in that folder, as shown in Figure 1-2.

All the folder contents must be pasted into the formatted micro SD card.

IMPORTANT: Do not simply copy the NOOBS_v2_7_0 folder itself. You must copy the folder *contents* onto the SD card or the card will not be bootable by the RasPi.

Writing the Raspbian Image to a Micro SD Card

Creating a Raspbian image is slightly different from the NOOBS process. In this case, you do not have to format the micro SD card prior to writing the image. That part of the process is automatically done for you by the application that writes the image to the card. You will need to set up an appropriate application based on your host computer. For a Windows machine, I highly recommend that you use

the Win32DiskImager available from https://sourceforge.net/projects/win32diskimager/files/latest/download.

The download is a Zip file, which will need to be extracted prior to use. All you need to do is run the application and select where the disk image is located, as well as the appropriate micro SD card logical file letter. Figure 1-3 shows my configuration screen for writing the Raspbian Stretch version to a micro SD card on a Windows machine.

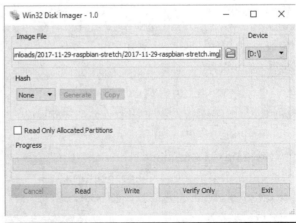

Figure 1-3   Win32DiskImager screenshot.

Figure 1-4 Etcher screenshot.

I recommend using the Etcher program if you are using a Mac to load the disk image. It is available from https://etcher.io/. This application functions in a very similar fashion to the Win32DiskImager program. Figure 1-4 is a screen shot of it being run on my MacBook Pro.

The next step in the setup process is to configure the image, once you have loaded it onto your micro SD card. The next two sections detail how to configure the NOOBS image first followed by the Raspbian image.

Configuring the NOOBS Image

The first step in configuring the NOOBS image is to set up a RasPi as a workstation. Do not attach the micro USB power supply to the RasPi before inserting the micro SD card holding the NOOBS file into the RasPi card holder. Ensure that the card is inserted *upside-down*, meaning that the printed side of the card is facing down. You ordinarily could not incorrectly insert the card unless you attempted to use unreasonable force, in which case you would likely break the holding mechanism.

Attach the USB power cable once the micro SD card is inserted, and you will see the initial screen, as shown in Figure 1-5.

Only one OS selection is shown in this particular downloaded image. Multiple OS selections were shown in earlier NOOBS

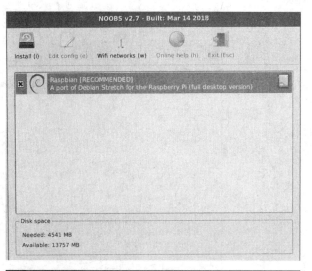

Figure 1-5 Initial NOOBS power-on screenshot.

versions, which leads me to believe that the Raspberry Pi Foundation folks have decided that the traditional Raspbian Debian distribution is their OS of choice. In any case, simply press the Enter key to fill in the checkbox next to the Raspbian selection. Next, you must press the "i" key to start the actual installation. Figure 1-6 shows the screen for the ongoing installation.

The whole installation process takes a while, depending mainly on the data throughput speed

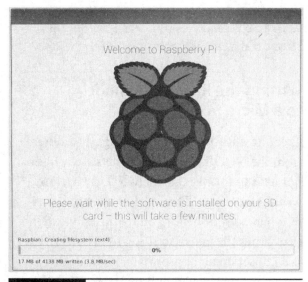

Figure 1-6 Raspbian installation screenshot.

Figure 1-7 Installation completion dialog box.

Figure 1-8 Starting the raspi-config utility.

of your micro SD card. This is why I strongly recommend that you use a Class 10 card to minimize tasks such as this installation. This installation took about 20 minutes when I did it. You should see the installation completion dialog box (Figure 1-7) after the process has completed.

The Raspbian OS has not yet been completely configured, even though the initial Raspbian

installation has finished. The OS must now be configured to your particular requirements, such as locale and network, using a useful utility named *raspi-config*. This utility is provided in the initial downloaded image. You run the raspi-config utility by opening a terminal window and entering the following command (as shown in Figure 1-8):

```
sudo raspi-config
```

Figure 1-9 shows the opening screen after this utility begins running.

There are nine selections, as you can see in the figure. Each one contains one or more configuration settings that you can use to address your requirements. In this initial

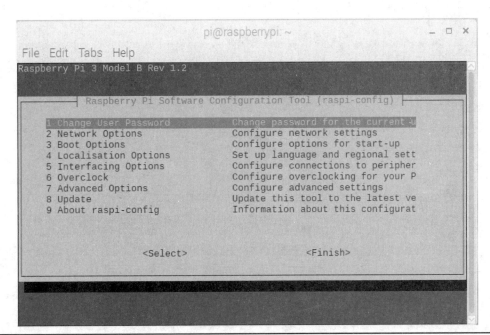

Figure 1-9 The raspi-config initial screenshot.

setup, I will be using three of the selections to configure the key settings for my configuration. I recommend that you initially duplicate these settings. However, you should feel comfortable in changing different settings as you gain additional expertise with the RasPi.

The first selection I chose was Localization Options. Figure 1-10 shows the new menu that appears when you select this option. The menu selections are straightforward, and they allow you to customize the RasPi configuration to suit your own country and keyboard setup. The keyboard configuration is particularly important because you will likely become frustrated trying to accomplish the WiFi configuration without changing the keyboard layout to match your country of origin.

The Interfacing Options menu is shown in Figure 1-11. This menu has eight selections, as shown in the figure. Which options you enable will depend on the types of devices you employ in your RaspPi system. I recommend enabling the following options to match the projects and procedures discussed in this book:

- Camera
- SSH
- SPI
- I2C
- Serial
- 1-Wire

You can easily add or subtract interfacing options at any time by rerunning the raspi-config utility. In any case, adding an interfacing option only minimally increases the size of the overall OS. Also note that enabling an interface only installs the associated driver(s) for that particular device. You will still need to install some additional application software to make a device fully functional within an HA system. I will discuss this application software at the appropriate time when dealing with a specific HA device.

The remaining step in the configuration process is to connect the RasPi to your home WiFi network. This is readily accomplished by modifying an existing configuration file to

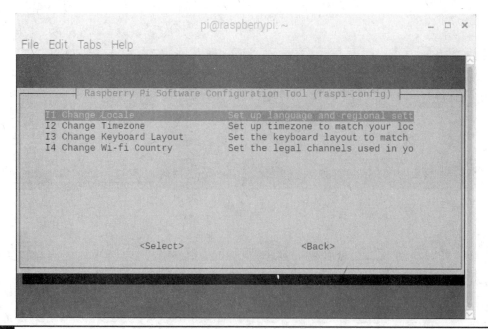

Figure 1-10   Localization Options menu selections.

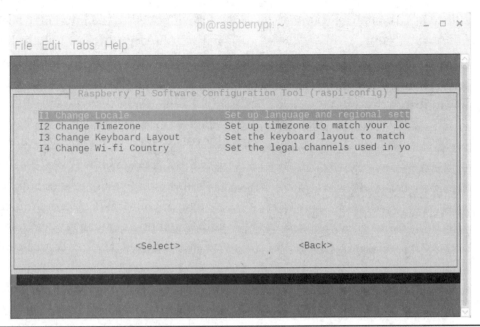

Figure 1-11 Interfacing Options menu selections.

connect the RasPi to your network. Open a new terminal window, and enter the following:

```
sudo nano /etc/wpa_supplicant/wpa_
supplicant.conf
```

You will then need to enter the following code snippet into the file following the last line in the file:

```
network={
    ssid="<your wifi name>"
    psk="<your wifi password>"
}
```

Figure 1-12 shows my modified file with the WiFi network name (ssid) and placeholder for the password (psk). Press CTRL-O and then ENTER

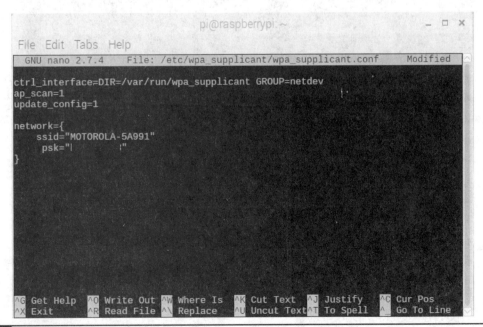

Figure 1-12 Modified wpa_supplicant.conf file.

to save the modified file. Then press CTRL-X to exit the nano editor application.

You need to reboot the RasPi to finish configuring the WiFi. Enter the following in the terminal window to reboot the computer:

```
sudo reboot
```

I recommend that you enter the following command into a terminal window to check whether the RasPi has connected to your home network:

```
ifconfig
```

Figure 1-13 shows the result of entering this command on my RasPi system.

You should be able to see in the wlan0 section that a local IP address of 192.168.0.6 was assigned to the RasPi by the home WiFi router. This assignment confirms that the RasPi is able

to be connected to the Internet. Check to see that your home router is set up for DHCP in case you do not see an IP address similar to the one shown in the figure. Also recheck that you entered the correct information into the wpa_supplicant.conf file. There is no error check for this edit, and the Raspbian OS will simply not connect if you have made an error in this edit.

At this point, you have successfully set up and configured your RasPi system. You next need to update and upgrade your system to ensure that the latest Raspbian OS software is installed.

Updating and Upgrading the Raspbian Distribution

The Raspbian Linux distribution is always being improved, as mentioned earlier. It is very easy to ensure that you have the latest updated and upgraded distribution once you have established

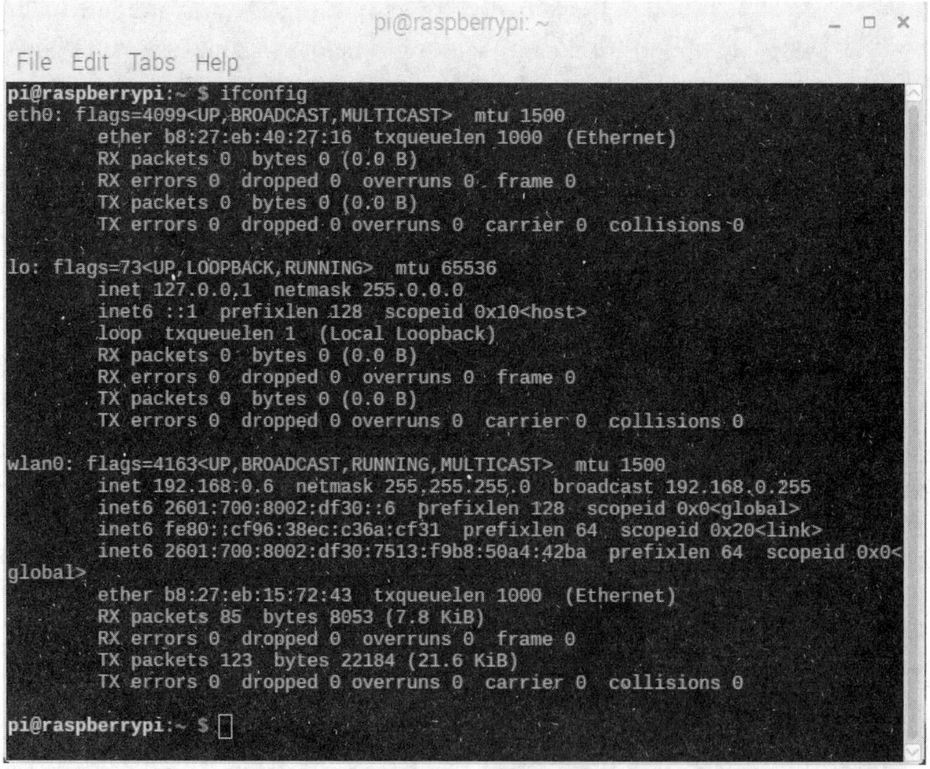

Figure 1-13 The ifconfig command display.

Internet connectivity. Enter the following command in a terminal window to update the installed OS:

```
sudo apt-get update
```

The update action changes the internal system's package list to match the current online package list. It does not actually change any of already installed packages if they are obsolete or outdated. Those changes are effected by entering the following command in a terminal window:

```
sudo apt-get upgrade
```

The update is reasonably quick if that original installed distribution is not too old. However, the upgrade action can take quite some time if a lot of outdated packages are already installed.

Just remember to always update prior to upgrading. All the projects in this book were created using an updated and upgraded Stretch Raspbian distribution. I have found that failing to update and upgrade can sometimes lead to some odd errors and system failures that are unexpected and puzzling.

You should have a completely functional RasPi system at this point in the installation and configuration process. The next section describes the general-purpose input/output (GPIO) portion of the RasPi, which is the primary means through which a RasPi can digitally control an HA system.

General-Purpose Input/Output

General-purpose input/output (GPIO) is the fundamental means by which a RasPi can either output or input a digital signal. A digital signal is any bistable signal, which is either on or off. The on state is typically represented by a 3.3-V level and the off state by a 0-V level. The 3.3-V level is the absolute maximum voltage that may be input into a RasPi without causing permanent damage. The actual voltage range at which a RasPi can sense an on or 1 signal is approximately 2.7 to 3.3 V. Similarly, the off or 0 state is approximately 0 to 0.2 V. Any voltage present on a GPIO input pin will sense a 0 with a voltage at or less than 0.2 V.

All the RasPi GPIO pins are part of the J8 header, which consists of a double row of 20 pins for a total of 40 pins. These pins and the associated GPIO designations are shown in Figure 1-14. This figure will be very useful to you when you attempt to duplicate this book's projects. You should note that a few additional pins on the header provide 3.3- or 5-V power as well as multiple ground connections. There are also two DNC pins, which is short for "Do Not Connect." In reality, the DNC designation only refers to the GPIO function. It turns out that

**Raspberry Pi 3 Model B
J8 GPIO Header**

	Pin No.		
3.3V	1	2	5V
GPIO2	3	4	5V
GPIO3	5	6	GND
GPIO4	7	8	GPIO14
GND	9	10	GPIO15
GPIO17	11	12	GPIO18
GPIO27	13	14	GND
GPIO22	15	16	GPIO23
3.3V	17	18	GPIO24
GPIO10	19	20	GND
GPIO9	21	22	GPIO25
GPIO11	23	24	GPIO8
GND	25	26	GPIO7
DNC	27	28	DNC
GPIO5	29	30	GND
GPIO6	31	32	GPIO12
GPIO13	33	34	GND
GPIO19	35	36	GPIO16
GPIO26	37	38	GPIO20
GND	39	40	GPIO21

Figure 1-14   GPIO header pin layout.

these two pins definitely have specific uses as alternate function pins, which I discuss later.

There are a total of 26 GPIO pins, which may be configured as either input or output at any given moment in time. A GPIO pin cannot simultaneously be both an input and an output. I have found that 26 GPIO pins are sufficient for any RasPi project that I have designed or duplicated. I believe that you will also come to the same conclusion.

One additional comment is necessary regarding how the GPIO pins are identified. You will notice in Figure 1-14 that all the GPIO pins are designated as "GPIOxx," where "xx" is a number ranging from 4 to 27. This identification is known as the *manufacturer's pin ID* or the *BCM mode* in Python programming terms, where BCM is the Broadcom manufacturer's abbreviation. Every pin in the figure also has a Pin No. designation, which is also called the *physical pin number*. Various pieces of device application software, which I mentioned earlier, sometimes use the physical pin IDs, whereas others may use the BCM designations. There is even another suite of application software named *wiringPi*, and it has its own unique pin IDs that are separate from the physical and BCM designations. This will become a very important issue in the next section, in which I provide a GPIO demonstration.

All this pin ID confusion is a natural consequence of open-source development, where there are really no enforced standards for such items as GPIO pin identification. You can rest assured that I will be quite clear regarding which GPIO pin ID to use in all my demonstrations and projects.

GPIO Demonstration

In this demonstration, I will show you how to control a LED using a single GPIO pin. This control action is made possible by an application named *wiringPi*, which provides a simple utility to turn on or off a LED directly connected to a GPIO pin. You will first need to install the wiringPi software, which is done as follows:

1. Install the git application. In a terminal window, enter the following:

   ```
   sudo apt-get install git
   ```

   NOTE: If you see any errors here, ensure that you have updated and upgraded the Raspbian distribution as discussed earlier. However, you may also discover that the git application has already been installed by viewing the following line:

   ```
   git is already the newest version
   (1:2.11.0-3+debian2)
   ```

2. Download wiringPi using git. Enter the following in a terminal window:

   ```
   git clone git://git.drogon.net/
   wiringPi
   ```

   NOTE: Ensure that you enter drogon and not dragon, a common mistake that I have previously made.

3. Build and install the wiringPi application. Enter the following commands in a terminal window:

   ```
   cd ~/wiringPi
   ./build
   ```

4. Check and test the wiringPi installation. Enter the following to check the installation:

   ```
   gpio -v
   gpio readall
   ```

Figure 1-15 shows the results of entering the last set of commands.

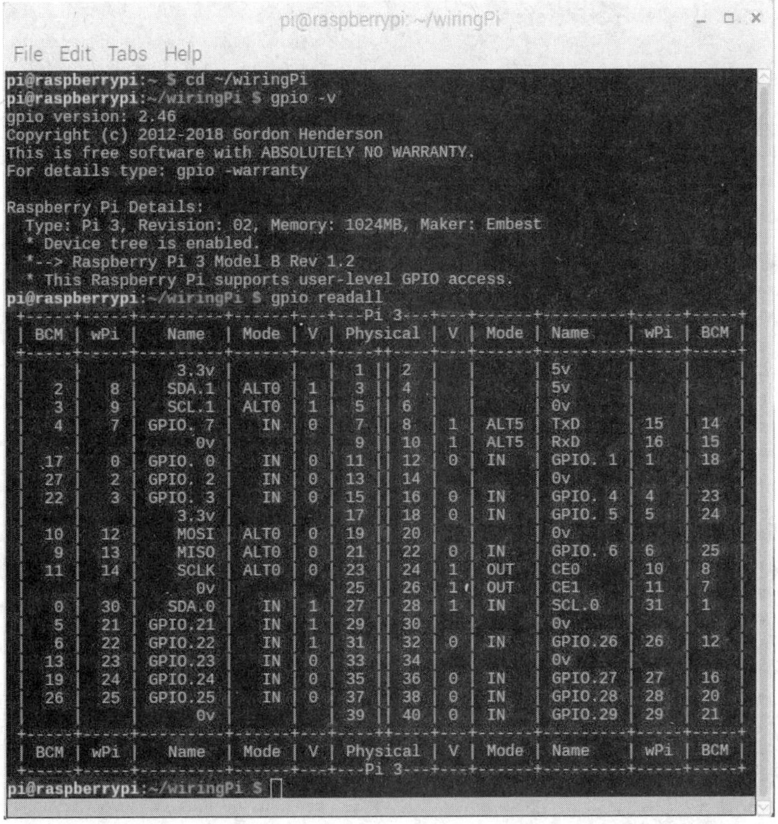

```
pi@raspberrypi:~ $ cd ~/wiringPi
pi@raspberrypi:~/wiringPi $ gpio -v
gpio version: 2.46
Copyright (c) 2012-2018 Gordon Henderson
This is free software with ABSOLUTELY NO WARRANTY.
For details type: gpio -warranty

Raspberry Pi Details:
  Type: Pi 3, Revision: 02, Memory: 1024MB, Maker: Embest
  * Device tree is enabled.
  *--> Raspberry Pi 3 Model B Rev 1.2
  * This Raspberry Pi supports user-level GPIO access.
pi@raspberrypi:~/wiringPi $ gpio readall
+-----+-----+---------+------+---+---Pi 3---+---+------+---------+-----+-----+
| BCM | wPi |   Name  | Mode | V | Physical | V | Mode | Name    | wPi | BCM |
+-----+-----+---------+------+---+----++----+---+------+---------+-----+-----+
|     |     |    3.3v |      |   |  1 || 2  |   |      | 5v      |     |     |
|   2 |   8 |   SDA.1 | ALT0 | 1 |  3 || 4  |   |      | 5v      |     |     |
|   3 |   9 |   SCL.1 | ALT0 | 1 |  5 || 6  |   |      | 0v      |     |     |
|   4 |   7 | GPIO. 7 |   IN | 0 |  7 || 8  | 1 | ALT5 | TxD     | 15  | 14  |
|     |     |      0v |      |   |  9 || 10 | 1 | ALT5 | RxD     | 16  | 15  |
|  17 |   0 | GPIO. 0 |   IN | 0 | 11 || 12 | 0 | IN   | GPIO. 1 | 1   | 18  |
|  27 |   2 | GPIO. 2 |   IN | 0 | 13 || 14 |   |      | 0v      |     |     |
|  22 |   3 | GPIO. 3 |   IN | 0 | 15 || 16 | 0 | IN   | GPIO. 4 | 4   | 23  |
|     |     |    3.3v |      |   | 17 || 18 | 0 | IN   | GPIO. 5 | 5   | 24  |
|  10 |  12 |    MOSI | ALT0 | 0 | 19 || 20 |   |      | 0v      |     |     |
|   9 |  13 |    MISO | ALT0 | 0 | 21 || 22 | 0 | IN   | GPIO. 6 | 6   | 25  |
|  11 |  14 |    SCLK | ALT0 | 0 | 23 || 24 | 1 | OUT  | CE0     | 10  | 8   |
|     |     |      0v |      |   | 25 || 26 | 1 | OUT  | CE1     | 11  | 7   |
|   0 |  30 |   SDA.0 |   IN | 1 | 27 || 28 | 1 | IN   | SCL.0   | 31  | 1   |
|   5 |  21 | GPIO.21 |   IN | 1 | 29 || 30 |   |      | 0v      |     |     |
|   6 |  22 | GPIO.22 |   IN | 1 | 31 || 32 | 0 | IN   | GPIO.26 | 26  | 12  |
|  13 |  23 | GPIO.23 |   IN | 0 | 33 || 34 |   |      | 0v      |     |     |
|  19 |  24 | GPIO.24 |   IN | 0 | 35 || 36 | 0 | IN   | GPIO.27 | 27  | 16  |
|  26 |  25 | GPIO.25 |   IN | 0 | 37 || 38 | 0 | IN   | GPIO.28 | 28  | 20  |
|     |     |      0v |      |   | 39 || 40 | 0 | IN   | GPIO.29 | 29  | 21  |
+-----+-----+---------+------+---+----++----+---+------+---------+-----+-----+
| BCM | wPi |   Name  | Mode | V | Physical | V | Mode | Name    | wPi | BCM |
+-----+-----+---------+------+---+---Pi 3---+---+------+---------+-----+-----+
pi@raspberrypi:~/wiringPi $
```

Figure 1-15 Screenshot for the check and test of the wiringPi installation.

The `gpio -v` command shows that wiringPi version 2.46 is installed and running. Your version may be slightly different depending on when you downloaded and installed the wiringPi application. What is more interesting is to view the results of the `gpio readall` command in Figure 1-15. The first thing you will probably notice is that there are different names for some of the GPIO pins than in the naming shown in Figure 1-14. This is so because many GPIO pins have alternate functions in addition to their default GPIO designations. For example, physical pins 3 and 5 are designated as GPIO2 and GPIO3 in Figure 1-14, whereas these same pins are designated as SDA.1 and SCL.1 in Figure 1-15. The designations SDA.1 and SCL.1 are for "serial data bus 1" and "serial clock 1," respectively. These control signals are associated with the I2C bit-serial protocol which is further discussed below along with some other supported protocols. The key point to remember is that alternative pin functions can be programmatically assigned to various GPIO pins to suit particular programming requirements. I recommend that you have both Figures 1-14 and 1-15 available when you are connecting devices to the GPIO header to help remind you of both the GPIO and alternate function designations.

There are also two columns in Figure 1-15 labeled "BCM" and "wPi" that are part of the confusing GPIO pin identification mentioned earlier. The BCM is the manufacturer's pin ID, whereas wPi is the corresponding wiringPi ID. You must use the wPi pin ID when programming a wiringPi application. Again, this situation frequently arises in open-source development, and you, as a competent developer, must adjust to the confusion.

Demonstration Hardware Setup

Figure 1-16 shows a schematic for controlling a LED using a single RasPi GPIO pin. The pin selected was wiringPi ID 0, which is also physical pin 11 and BCM ID GPIO17.

The series 330-Ω resistor shown in the figure is used to limit the current supplied by the GPIO pin. In this case, the current draw would be approximately 8 mA, assuming that the LED has a 0.7-V drop across it when forward biased. This current draw is well within the 15-mA maximum specification limit for a RasPi GPIO pin. The LED cathode is also connected to ground to complete the circuit.

Figure 1-17 is a Fritzing diagram that shows the main components of the circuit, including a T-Cobbler, which consists of a flat ribbon cable with one end terminated with a double-row header socket that plugs into the RasPi GPIO

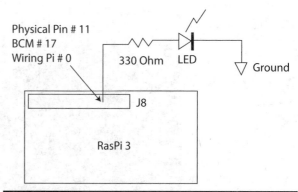

Figure 1-16   Demonstration circuit schematic.

header. The other end has a labeled plug that fits into a standard solderless breadboard, as shown in the figure. I have also included Figure 1-18, which is photograph of the physical components that make up the demonstration circuit.

The last three figures were all provided to clearly show how to set up a RasPi project. Later book projects may have a schematic, a Fritzing

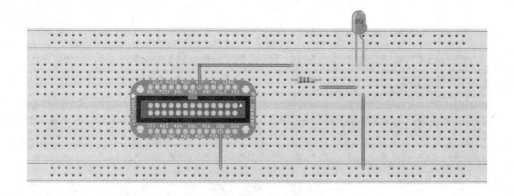

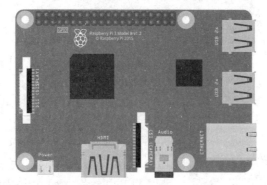

Figure 1-17   Demonstration circuit Fritzing diagram.

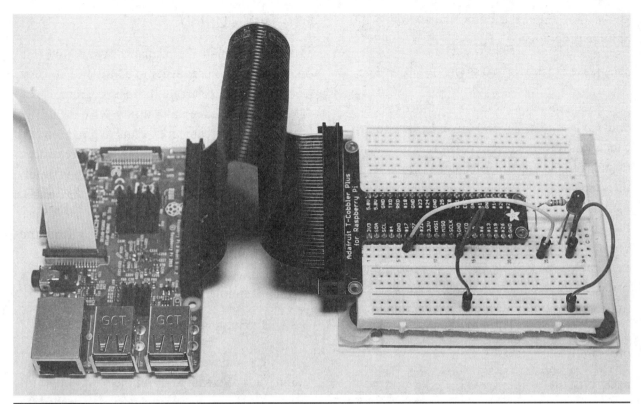

Figure 1-18 Actual demonstration circuit.

diagram, and/or a photograph depending on project complexity.

Next is a detailed discussion of the software required to enable the demonstration hardware.

Demonstration Software Setup

The wiringPi application requires the use of C language programs to control the GPIO pins. The demonstration program is a classic version of the "Hello World" program, which blinks a LED once per second. The following program listing should be input into a file named *blink.c* using the nano editor:

```
#include <wiringPi.h>
int main (void)
{
  wiringPiSetup ();
  pinMode (0, OUTPUT);
  for (;;)
```

```
  {
    digitalWrite (0, HIGH);
    delay(500);
    digitalWrite (0,  LOW);
    delay (500);
  }
  return 0 ;
}
```

You can start the nano editor by entering the following command in a terminal window:

```
sudo nano blink.c
```

Enter all the code as shown in the listing, and press CTRL-O to save the code in a file named *blink.c*. You will next need to press CTRL-X to exit the nano editor.

The code in the blink.c file now needs to be compiled and linked to form an executable application. Enter the following to compile and link the program:

```
gcc -Wall -o blink blink.c -lwiringPi
```

The `gcc` portion of the command invokes the Gnu C/C++ compiler and linker application, which was downloaded as part of the original OS image. The executable is named *blink* because that name immediately follows the `-o` option. And finally, the wiringPi library is linked with the compiled source code by the `-lwiringPi` option. The completely compiled and linked executable is named *blink* and will be found in the home directory if you haven't changed it from when you ran the preceding command.

Next, enter the following command to run the program:

```
sudo ./blink
```

If all went well, you should now observe the LED blinking once per second. If you do not see it blinking, then I recommend the following:

- Recheck the LED circuit, and confirm that the correct GPIO pin is connected.

- Recheck the LED orientation. It won't light if it is connected in reverse.

- Confirm that the program has the correct wiringPi pin ID. The program may have compiled and linked properly, but if the pin ID is incorrect, it cannot function as expected.

The LED will continue to blink indefinitely because the GPIO control statements are within a "forever" loop created by the `for(;;)` statement. You will need to press CTRL-C to stop the program and LED from blinking. The CTRL-C key combination is known as a *keyboard interrupt* and is quite useful to stop a program and/or process that is currently running on a RasPi.

The Make Utility

A handy alternative is available if you don't like or desire to use the nano editor and gcc application to create the test program. This alternate is the `make` utility, which will automatically create an executable program using a preset script and premade source code. The script is found in a *makefile*, which is simply a bundle of OS-level commands that the `make` utility follows to create an executable file. To use `make`, you must first change directories from the home directory to the directory that holds the example programs in the wiringPi directory. The terminal command to make this change is

```
cd ~/wiringPi/examples
```

Then all you have to do enter these next commands to create and run the blink program:

```
make blink
sudo ./blink
```

The LED should start blinking once per second after these commands are executed. The blink.c file stored in the Examples directory is almost identical to the program listing provided previously in this chapter.

There are other programs in the Examples directory that control additional LEDs, such as blink8.c and blink12.c, which blink the first 8 and 12 GPIO pins, respectively. There are also a series of programs that control various devices that are not used in this book but could be explored on your own if you are so interested.

Natural Human Interaction

Natural human interaction (NHI) is the name given by Google's AI scientists and researchers to the new AI field of how computers can interact with humans in the same way as humans interact with other humans. Other AI researchers have named this field the *natural user interface* (NUI), but it all relates to simplify how a human can control a computer system using only natural behaviors such as speech, hand motions, hearing, or even having the computer recognize a facial image. Modern HA systems must include NHI features in order to be attractive to potential nontechnical users so as to be viable and have widespread use. I focus on speech recognition as well as computerized auditory response as the appropriate technologies to use in this book.

Google, in cooperation with the *MagPi* magazine publisher, has made available a project kit that allows makers to experiment with NHI. This inexpensive kit is named the *Google AIY Voice Kit for the Raspberry Pi*, for which I have provided a source in this chapter's Parts List.

A comprehensive booklet included with the kit that describes in detail how to set up both the physical device and the software required to operate it. You will also need to download a complete AIY Voice image in a manner similar to the procedure described previously. This image is available from the following link: https://aiyprojects.withgoogle.com/voice -assembly-guide-1-get-the-voice-kit-sd-image. The kit build is very easy, requiring no tools other than a 00 Phillips head screwdriver. Figure 1-19 shows the completed kit with a RasPi 3 already set up inside the cardboard box enclosure.

The kit has only four electronic components, which are depicted in the block diagram in Figure 1-20. I will first explain how the HAT module, microphone array, and LED/ pushbutton components function before detailing how to install the RasPi software that will enable the voice kit itself.

Figure 1-19 Google AIY Voice Kit for the Raspberry Pi.

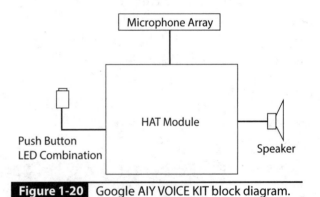

Figure 1-20 Google AIY VOICE KIT block diagram.

HAT Module

The Hardware Attached on Top (HAT) module is the key component in the voice kit. It contains all the interface and sensor processing capabilities that provide for both speech recognition and audio output. It also contains some additional features that will enable you to control some additional external devices via some RasPi GPIO pins. The HAT module is based on a recent Raspberry Pi Foundation standard that specifies how third-party manufacturers may design daughter boards that plug into a RasPi GPIO header to provide extended capabilities. The HAT standard provides for both physical dimensions and embedded software interface specifications. The typical way the embedded software interface is implemented is through the use of a serial *electrically erasable programmable read-only memory* (EEPROM) chip. The voice HAT module uses a 4-kB serial EEPROM that can be accessed by the connected RasPi to provide it with all the necessary board configuration parameters required for the voice interface application. All the power normally required for the voice HAT module is provided by the GPIO header power pins. This HAT module does have a provision to connect an external power source in case there is a requirement to power connected devices that require power in excess of what can be supplied through the header pins. There is no additional general-purpose memory

on the board because that is provided by the connected RasPi. Figure 1-21 shows the voice HAT module.

Six servo connections are also provided on the HAT module. These connections allow you to directly control low-voltage devices such as servo motors using six associated RasPi GPIO pins. The module's servo connections are shown in Figure 1-22.

These connections are printed-circuit board (PCB) solder pads to which you can easily solder three-pin headers to allow easily connection of standard hobby-grade servo motors and other similar devices. Table 1-1 details how the RasPi GPIO pins correspond to the voice HAT servo connections. You should note there is a series 220-Ω resistor connected that acts as a current limiter for the connected GPIO pin.

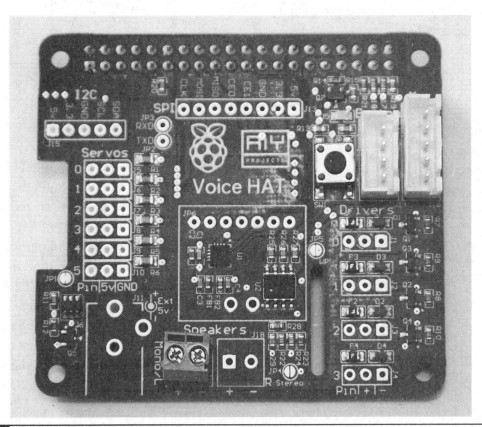

Figure 1-21 Voice HAT module.

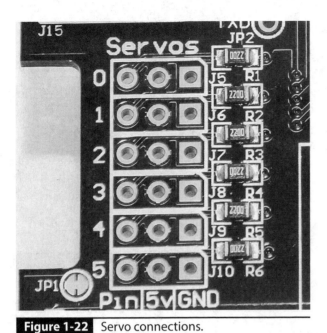

Figure 1-22 Servo connections.

TABLE 1-1 Voice HAT Servo to RasPi GPIO Pin Connections

Servo Pin Number	RasPi Physical Pin Number	RasPi BCM Pin Number
0	37	26
1	31	6
2	33	13
3	29	5
4	32	12
5	18	24

In addition, there are four sets of driver connections provided on the HAT board. These connections are shown in Figure 1-23. They are labeled "Driver 0" through "Driver 3" and are similar to the servo connections except that the RasPi GPIO pins directly control a field-effect transistor (FET), which, in turn, controls the current flow to the PCB solder pads shown in the figure. This configuration will allow for up to 0.5 A to be supplied, far in excess of what is available through a normal RasPi GPIO pin. Figure 1-24 shows the schematic for one driver channel. Notice in the schematic that each driver channel has a 500-mA poly fuse in series with the FET drain to prevent an overload and

Figure 1-23 Driver connections.

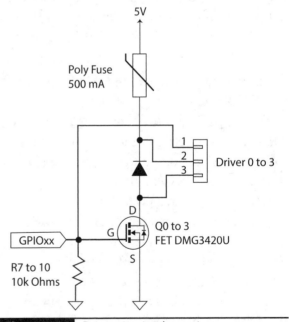

Figure 1-24 Driver circuit schematic.

subsequent permanent damage to the driver circuit. The poly fuse is resettable, meaning that it will start conducting again after cooling down when subjected to a transient overload condition.

Table 1-2 details the connections between the RasPi GPIO pins and the driver channels.

TABLE 1-2 Voice HAT Driver Channel to RasPi GPIO Pin Connections

Driver Channel Number	RasPi Physical Pin Number	RasPi BCM Pin Number
0	7	4
1	11	17
2	13	27
3	15	22

The question that might naturally arise for you is when should you use a driver channel. The answer would be that you use one to handle any relatively high current load that is beyond the normal RasPi GPIO current limit of approximately 20 mA. Such loads might consist of relay coils, which are often used in HA projects. The relays themselves then could handle much heavier currents, including mains voltage and current, for such things as appliances and lighting fixtures.

There a few more items on the HAT board, which I have lumped together in this next brief discussion. A digital-to-analog converter (DAC) combined with a class D audio amplifier takes the converted analog audio stream coming from the RasPi I2S interface and sends it to a 4-inch loudspeaker. I can readily attest that the speaker is quite loud and remarkably clear. Next is a power supply that can take an external input from a barrel connector (not supplied with the kit) and generate a regulated 5-V power source for both the servo and driver outputs. Use of this supply is absolutely required if you want to power servos and/or driver channel devices. The organic RasPi power supplied through the GPIO pins is not capable of meeting the heavy current demands, and you will most likely cause permanent damage to the RasPi if you attempt to operate in such a fashion.

Table 1-3 is a compilation of bit-serial communication interface pins that are available using PCB pads and a five-pin header located on the HAT module. The interfaces consist of

- Serial peripheral interface (SPI)
- Interintegrated circuit (I2C)
- Universal asynchronous receiver transmitter (UART)

TABLE 1-3 Bit-Serial Communication Interface Pin Connections

Pad Label	Bit-Serial Interface	Pin/Pad Label	Remarks
JI3	SPI	5 V	—
JI3	SPI	3.3 V	—
JI3	SPI	GND	—
JI3	SPI	CE1	Chip enable 1
JI3	SPI	CE0	Chip enable 0
JI3	SPI	MISO	Master in, slave out
JI3	SPI	MOSI	Master out, slave in
JI3	SPI	CLK	Clock
JI5	I2C	SDA	Serial data
JI5	I2C	SCL	Serial clock
JI5	I2C	GND	—
JI5	I2C	3.3 V	—
JI5	I2C	5 V	—
JP2	UART	TXD	Transmit out
JP3	UART	RXD	Receive in

These pins can be useful for connecting additional modules or devices to use with the HAT, as well as an easy access point for the bit-serial signals that otherwise would be unavailable because the HAT mode covers the entire RasPi GPIO header. These pins are clearly silk-screened on the HAT and can be seen in Figure 1-25.

Figure 1-25 Voice HAT bit-serial communication interface pads and pins.

Figure 1-26 Dual microphone array.

Microphone Array Module

The voice kit employs a dual microphone array to sense and send digitized user voice messages to the RasPi. The digitized signals are in the I2S Bit-Serial Protocol, which is the same protocol used to send digital announcements to the DAC-amplifier combination. Figure 1-26 shows the microphone array, in which you can clearly see the digital microphones located at each end of the slender PCB module.

Each microphone is a Knowles MEMS unit, Model SPH0645LM4H. An uncased microphone is shown in Figure 1-27 alongside a block diagram for the unit.

Any incoming sound wave will be converted from its analog form to an equivalent digital data stream by the microphone unit using the I2S Bit-Serial Protocol, with the data subsequently sent over to the RasPi. Each

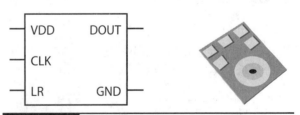

Figure 1-27 Knowles digital microphone.

microphone can be selectively enabled depending on whether or not a solder bridge (JP1, JP2) is in place. The default configuration is to have both microphones enabled.

Pushbutton-LED Combination

There is also a relatively large pushbutton-LED combination device, as shown in Figure 1-28. The purpose of this device is to activate the RasPi/HAT system to begin recognizing a user's voice and to light a LED to provide feedback to the user that the system is enabled. The button's

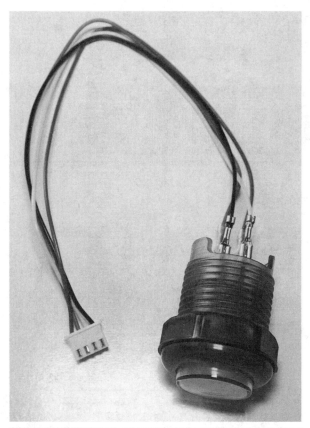

Figure 1-28   Pushbutton-LED combination.

digital signal is connected to GPIO physical pin 16 (BCM #23), and the LED is connected to GPIO physical pin 22 (BCM #25).

Google Voice Assistant Software Installation

This section discusses in detail the key steps involved in the installation of the Google Voice Assistant software. Step-by-step instructions are also provided in the instruction pamphlet included with the Google Voice Kit. The instructions included in this section presume that you already have downloaded and created a micro SD card from the AIY Voice image mentioned earlier. This image contains several Python script directories, which allow the RasPi to connect with the Google server and

implement both voice-recognition and speech-generation functions. These steps should be accomplished using the RasPi's Web browser.

Create a Google Account

The very first action you must take is to create a Google account. I am pretty sure that most of my readers already have a Google account, but if you don't, just go to https://accounts.google.com/signup. This is a very simple action, but it is absolutely required before proceeding to the next step.

Log into the Google Cloud

The next step is to log into the Google cloud (GC) website with your Google account user name and password. Normally, your regular email account name is the user name, and whatever you set for your Google account password is the same for the cloud login. The GC login link is https://console.cloud.google.com.

Create a Google Cloud Project

You will need to create a Google cloud project (GCP) in order to continue with the project. Just click on the GCP icon in the upper left-hand corner of the cloud console webpage. Figure 1-29 shows the result of this action. You will see that I already created a project (as shown in the figure). Yours will be blank when you first open it. Click on the **+** icon to open the dialog to name your project. I suggest naming the project *Voice Assistant*, just as I did in the figure.

Enable the Google Assistant API

Click on the triple-bar icon in the upper left-hand corner of the GCP menu bar, and then click on the Library selection from the drop-

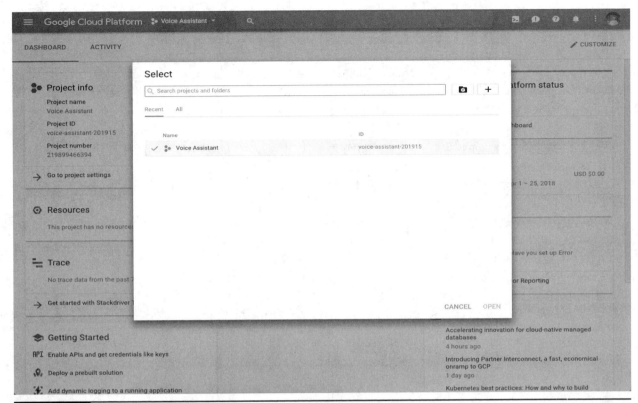

Figure 1-29 New GCP dialog screen.

down menu. Then enter "Google Assistant API" into the search text box. Next, click the Enable icon to include the Google Assistant API in your project.

The Google Assistant API uses the gRPC framework to implement the Voice Assistant functions between the Python client, which is stored on the RasPi, and the Google server software, which is located at a remote Google data service center. I have included a sidebar, "Google Remote Procedure Call," that delves into the gRPC framework to further expand your knowledge regarding this important technology.

Obtain Credentials

Security should always be a consideration in any computer project, especially ones that use the Web. This Voice Assistant project is

vulnerable to being hacked, which is why there is strong security built into this Web application. This security is implemented using OAuth 2.0 authentication on both the client and the server. OAuth 2.0 is a token-based authentication sequence. It is implemented in a process clearly shown in Figure 1-30, which is a data-sequence flow diagram showing both the client and server as they request/respond to each other in creating a secure link.

You will need to create an OAuth 2.0 client by first clicking on APIs & Services on the cloud console screen. Then select Credentials, and click on Create Credentials when the Credentials screen appears. You will first need to configure your consent by clicking on the Configure consent screen. Choose an appropriate product name, and then click Save. I recommend the name *Voice Assistant*.

Google Remote Procedure Call

gRPC is an abbreviation for "Google Remote Procedure Call," where a client application can directly call methods on a remote machine as if it were a local object. This design helps you to easily build distributed applications and services. The gRPC framework defines what services are available, how to request those services, and finally, how to deal with the returned objects. The server side is responsible for implementing the entire interface and properly handling all client calls. The client side, which is the RasPi in our case, contains a stub that mirrors the same exact methods that are implemented on the server side. The gRPC is set up to handle a variety of languages. Python was chosen to be the language used for the client-side implementation for the Voice Assistant. I am not sure what language the Google server side uses, but I suspect that it likely is C/C++ for efficiency.

Data streams are exchanged between the client and server using protocol buffers. This is an open-source implementation where the serial data to be exchanged are set up in a PROTO file, which is a text file that uses a .proto extension. The protocol-buffered data are structured as messages, where each message is a record containing name/value pairs. These pairs are also called *fields*. The following is a simple record example:

```
message Person {
    string name = 1;
    int32 id = 2;
    bool has_auth = 3;
}
```

A buffer protocol compiler is used once all the data structures have been defined. The compiler will generate appropriate language-specific classes from the PROTO definitions. In addition, unique accessor methods will be created for each of the mutable variables. Methods are also provided that will serialize and parse the raw data bytes to language-appropriate objects.

gRPC services are defined in PROTO files with RPC parameters and return types specified in accordance with protocol buffer messages. The following is an example of a service definition:

```
// The greetings service definition
service GreetingService {
    // Define a RPC operation
    rpc greeting(HelloRequest) returns(HelloReply);
}

// The request message has the payload; user's name
message HelloRequest {
    string name = 1;
}

// The response message containing the greetings
message HelloReply {
    string message = 1;
}
```

There have been several revisions to the protocol buffer, with the latest being proto3, which I believe is used with gRPC. The latest documentation for gRPC protocol buffers is available from https://developers.google.com/protocol-buffers/docs/reference/overview.

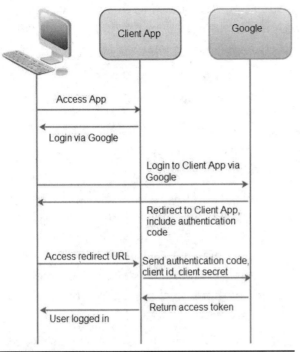

Figure 1-30 OAuth 2.0 data-sequence flow diagram.

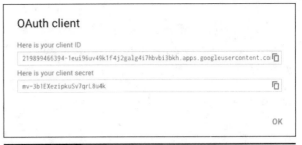

Figure 1-31 Client ID and client secret.

Click on the Create button. The program will briefly pause while the client ID and client secret are both generated. Figure 1-31 shows a sample screen shot of a client id and client secret.

Don't worry about copying the obscure data, they will be automatically sent at the appropriate time according to the sequence diagram I presented earlier.

Download and Rename Credentials

You will need the new credentials to be stored in a local RasPi file. Select the Voice Recognizer credential shown on the Credentials screen. Then click on the Download JSON icon, as shown in Figure 1-32. This action will download a JSON file containing both the client ID and client secret to a file in your Downloads directory. The default file name is very complex, with many

Next, you need to name your newly created credential. Go back to the Credentials screen, and click on the Create Credentials button. Select OAuth client ID from the drop-down menu. Click on the radio button "Other" for the application type. Enter a name in the textbox that pops up. I suggest using the name *Voice Recognizer*, which fits the intended application.

Figure 1-32 Download credentials screenshot.

numbers and letters. You will need to rename it to make it practical to access. To do this, open a terminal window and enter the following commands:

```
cd Downloads
ls
mv client_secret
```

Now press the TAB key, which will automatically fill in the rest of the download file's letters and numbers. Append the following to the command you are building in the terminal window:

```
/home/pi/assistant.json
```

Now press the ENTER key, and you will have a new JSON file named `assistant.json` in the home directory, which is required for the following steps.

Activity Controls

Activity controls add functionality to your Google account. In this case, you will be enabling the Google Voice & Audio Activity control for your account. By doing so, you will be enabling your account to

- Learn the sound of your voice
- Learn how you say words and phrases
- Recognize when you say "OK Google"
- Improve speech recognition across Google products that use your voice

Google saves your voice recordings and other audio present when you activate the control by saying "OK Google." Activation also can be initiated by pressing a pushbutton if the system is so designed with such a device. The voice kit uses the pushbutton approach to activate voice recordings.

Go to the following link to activate the Voice & Audio Activity control: https://myaccount .google.com/activitycontrols/audio. Enable the activity by clicking on the slider shown in Figure 1-33.

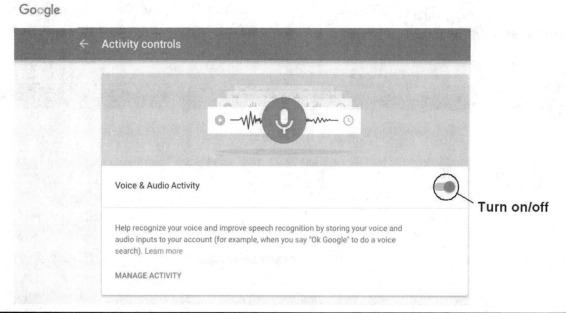

Figure 1-33 Enabling the Voice & Audio Activity control.

Figure 1-34 Terminal window log for a voice command.

Note that voice inputs cannot be saved if the Voice & Audio Activity control is turned off, even if you are signed into your Google account.

Testing the Voice Kit

You are now ready to test the voice, provided that you have completed all the previous steps. Enter the following commands in a terminal window to start the voice kit test:

```
cd AIY-voice-kit-python
src/examples/voice/assistant_grpc_demo.py
```

Next, press the large pushbutton, which should start glowing green. Then clearly speak your voice command. My first command was

"What time is it?" The voice kit clearly and loudly responded with the local time. You should try out a bunch of different commands to see how Google responds. Enter the CTRL-C key combination to stop the program. Figure 1-34 shows the terminal window log of how the system responds to parsing the voice command.

The following is the program listing for the demonstration Python program. I have included it here to illustrate how relatively simple it is to implement voice recognition and speech output using Google's client/server approach. This listing will also be used for some simple modifications that will be discussed in the next section.

```
#!/usr/bin/env python3
# Copyright 2017 Google Inc.
# Licensed under the Apache License, Version 2.0 (the "License");
# you may not use this file except in compliance with the License.
# You may obtain a copy of the License at
# http://www.apache.org/licenses/LICENSE-2.0
# Unless required by applicable law or agreed to in writing, software
```

```
# distributed under the License is distributed on an "AS IS" BASIS,
# WITHOUT WARRANTIES OR CONDITIONS OF ANY KIND, either expressed or
# implied.
# See the License for the specific language governing permissions and
# limitations under the License.

"""A demo of the Google Assistant GRPC recognizer."""
import logging
import aiy.assistant.grpc
import aiy.audio
import aiy.voicehat
logging.basicConfig(
    level=logging.INFO,
    format="[%(asctime)s] %(levelname)s:%(name)s:%(message)s"
)
def main():
    status_ui = aiy.voicehat.get_status_ui()
    status_ui.status('starting')
    assistant = aiy.assistant.grpc.get_assistant()
    button = aiy.voicehat.get_button()
    with aiy.audio.get_recorder():
        while True:
            status_ui.status('ready')
            print('Press the button and speak')
            button.wait_for_press()
            status_ui.status('listening')
            print('Listening...')
            text, audio = assistant.recognize()
            if text:
                if text == 'goodbye':
                    status_ui.status('stopping')
                    print('Bye!')
                    break
                print('You said "', text, '"')
            if audio:
                aiy.audio.play_audio(audio)
if __name__ == '__main__':
    main()
```

Extending the Voice Kit Functionality

This section shows you how to increase the voice kit functions using a modification to the demonstration program previously listed. The first thing you should do is to create a copy of the original program, on which any modifications will be applied. This action ensures that an unmodified program listing is preserved so that you can always restore the original program functions. Always make modifications to the copy, never to the original. Enter the following commands in a terminal window to make the first program copy:

```
cd ~
cd AIY-voice-kit-python/src/examples/voice
cp assistant_grpc_demo.py demo1.py
sudo nano demo1.py
```

Note that I used *demoN.py* as the copy name, where N starts at 1 and progresses in a sequence. You should also make a list of the *demoN.py* names and what modifications have been done to them. It is very easy to get lost in program modifications without careful attention to what changes have been made to what programs.

This program modification will be to turn off the RasPi by using the voice command "Power down." Use the nano editor to insert the following additional code, which is in bold text, as shown in the listing:

```
"""A demo of the Google Assistant GRPC recognizer."""
import logging
import aiy.assistant.grpc
import aiy.audio
import aiy.voicehat
import subprocess
logging.basicConfig(
    level=logging.INFO,
    format="[%(asctime)s] %(levelname)s:%(name)s:%(message)s"
)
def shutit():
    subprocess.call('sudo shutdown now', shell=True)
def main():
    status_ui = aiy.voicehat.get_status_ui()
    status_ui.status('starting')
    assistant = aiy.assistant.grpc.get_assistant()
    button = aiy.voicehat.get_button()
    with aiy.audio.get_recorder():
        while True:
            status_ui.status('ready')
            print('Press the button and speak')
            button.wait_for_press()
            status_ui.status('listening')
            print('Listening...')
            text, audio = assistant.recognize()
            if text:
                if text == 'goodbye':
                    status_ui.status('stopping')
                    print('Bye!')
                    break
                if text == 'power down':
                    aiy.audio.say('Shutting Down!')
                    shutit()
                print('You said "', text, '"')
            if audio:
                aiy.audio.play_audio(audio)
if __name__ == '__main__':
    main()
```

I made the code modifications and entered the following commands to test the program:

```
cd ~
cd AIY-voice-kit-python
src/examples/voice/demo1.py
```

I observed that the RasPI did shut down shortly after I voiced the command "Power down." This program modification should clearly show that it is relatively easy to extend the RasPi voice interface functions. I will also demonstrate in some of the following chapters how to directly control GPIO pins using voice commands. This feature will be very important in implementing a variety of HA functions.

Summary

This chapter started with a brief discussion regarding several HA design approaches. These included using a natural human interaction (NHI) function such as voice activation.

The next sections concerned how to set up a Raspberry Pi (RasPi) as an HA microcontroller. Topics included

- Creating an OS disk image on to a micro SD card

- Raspbian configuration
- Updating and upgrading the OS

I next presented a comprehensive discussion on the general purpose input/output (GPIO) system. GPIO pins are a very important design feature that provides a means for the RasPi to control HA interfaces. I also demonstrated a simple LED blink program, which used the wiringPi library for GPIO pin control.

Several sections followed concerning how to build and program the Google AIY Voice Kit. This kit was used to demonstrate a voice-recognition capability using a RasPi. I explained how each principal kit component worked, including the Hardware Attached on Top (HAT) module.

A detailed account of how to set up a Google developer's account followed. This account is necessary to access the Google cloud platform, which services the voice kit for both voice recognition and speech output.

The chapter concluded with a demonstration of how to modify the default voice kit program such that the RasPi shuts down on recognizing the voice command "Power down."

## CHAPTER 2

Interfacing a Google Home Device with a Raspberry Pi

THIS CHAPTER INCLUDES a detailed discussion of how Google's Home device works and how it interacts with Google's Web servers to provide intelligent home-based voice services. The overall Google service is called *Google Assistant* and is also the same service described in Chapter 1 for the Google Voice Kit. This chapter also has a comprehensive discussion of how to connect a RasPi to a Home device to take advantage of the professional voice-recognition and speech-generation functions. This approach is significantly different from the one taken in Chapter 1, where a Google Voice Kit was connected to a RasPi to enable fairly basic voice recognition and speech generation. That approach required the RasPi to perform many functions that are done solely within the Home device. This chapter's approach is to have the Home device function more as a coprocessor than as a RasPi peripheral. This means that the Home device offloads a substantial amount of the computational burden from the RasPi, which, in turn, permits the RasPi to have more computational assets available to efficiently perform other system functions.

Google Home Device

It is important to have a detailed discussion regarding the Google Home device before delving into how it functions with a RasPi. Figure 2-1 shows the device, which is self-contained and connects to the Internet through a local WiFi link.

Parts List

Item	Model	Quantity	Source
RasPi 3	B or B+	1	adafruit.com amazon.com mcmelectronics.com
Google Home device	Home	1	amazon.com
Power Switch Tail	II	1	amazon.com
Qiachip 433-megahertz (MHz) transmitter/receiver, four-channel	—	1	amazon.com
Alternating-current (AC) table lamp	Commodity	1	Various

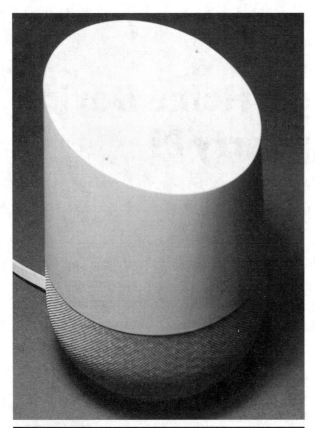

Figure 2-1   Google Home device.

Some key technical specifications are as follows:

- Dimensions: 3.79 inches (96.4 millimeters [mm]) in diameter by 5.62 inches (142.8 mm) high

- Weight: 1.05 pounds (477 grams [g])

- Colors: White with slate fabric base (standard); other colors available

- Supported audio formats: HE-AAC, LC-AAC, MP3, Vorbis, WAV (LPCM), Opus, and FLAC (hi-res 24 bits, 96 kilohertz [kHz])

- Wireless communications: 802.11b/g/n/ac (2.4 GHz/5 GHz) WiFi, Bluetooth 4.1, and Near-Field Communication (NFC)

- Speakers: 2-inch active driver with dual 2-inch passive radiators

- Far-field voice recognition

- Power supply: 16.5-V, 2-A wall adapter

- Ports and connectors: Direct-current (DC) power jack

- Supported operating systems: Android (4.4 and higher) and iOS (9.1 and higher)

Please note that only the WiFi communication link is enabled in the default Google Home configuration. The internal communications module shown as part of the Figure 2-2 block diagram is capable of supporting Bluetooth and NFC as well as the default WiFi protocol.

A lot of hardware is contained in the device, which makes it very user friendly, including a touch interface, LED displays, and the voice interface. The top of the device contains a grid of capacitive pads, which can sense a user's finger touch and react appropriately. The top also has an embedded multicolored 12-LED ring array, which provides some visual feedback to the user. Finally, a two-element *micro-electro-mechanical systems* (MEMS) microphone array is mounted on the top surface, which senses the user's voice. If you carefully examine Figure 2-1, you can see the two tiny holes for the microphones. It is very important to keep these holes clear of any blocking material or you will degrade or even stop the device from functioning properly.

This device consumes a fair amount of power, which may be inferred from the 16.5-V, 2-A power supply. This suggests that it would be impractical to consider using this device in a portable, battery-powered project. The power consumption is not an important limitation because the device will likely be used in a static configuration in the user's home while constantly being plugged into the AC mains.

The device uses two microprocessors, one of which I labeled the "auxiliary processor" in Figure 2-2, which controls and processes the user's interactions. The other microprocessor, which I labeled the "main processor" in Figure

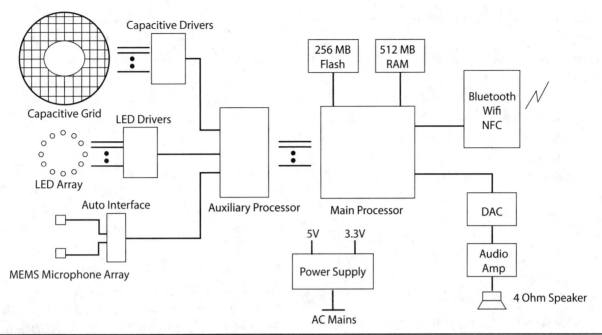

Figure 2-2   Google Home block diagram.

2-2 communicates with the Google Assistant service, processes audio streams, and does several other related functions. The main processor has a 256-megabyte (MB) Flash memory, which will automatically be updated to the latest version during your initial setup. It also has 512 MB of RAM, which holds the dynamic data required to support real-time operations.

The device uses a 2-inch active 4-Ω speaker along with two 2-inch passive radiators to produce high-quality audio. It can potentially generate such a high volume that you will want to turn it down so that it will not create too much of a disturbance.

Configuring the Google Home Device

The device must be configured initially using a smartphone app. You must first load the app named *Google Home* either on an Apple (iOS) or Android smartphone or equivalent tablet device. Then follow the on-screen instructions to enable the device. I used the iOS version without any problems and promptly got the device up and

running correctly. I am refraining from including any instructions regarding how to configure the device because the step-by-step procedures displayed on the smartphone or tablet are abundantly clear.

I would strongly suggest that you take some time to learn how to use the device in a stand-alone fashion. I believe that you will be impressed with its inherent capabilities and functionalities. The more you understand how to successfully use the device, the easier it will be for you to understand how to use it with a RasPi.

Connecting a RasPi to the Google Home Device

The first thing you must realize is that no direct physical connection can be easily made between a Home device and a RasPi. Instead, a logical connection between the two must be made using the Cloud. In this way, the Home device connects with a Web-based server that in turn connects with another Web service that sends commands to yet another Web server hosted on the RasPi. It turns out that there are a number of approaches to implementing the logical connection. I have taken an approach using software and protocols that are relatively uncomplicated and easy to use and with which I was already familiar. I will create a very simple demonstration project to show you how this all works in a step-by-step manner. The project will be to remotely control a single LED connected to a GPIO pin on the RasPi. The LED will be turned on by the phrase "OK Google, turn on the LED" and turned off by the phrase "OK Google, turn off the LED."

The first element of the project concerns how the Home device initiates a Web request that is eventually sent to the RasPi's Web server.

ifttt.com

This section's heading is the URL for the Web service that enables a Google Home device to send a Web request to a RasPi-hosted Web server. The URL name is short for "if this then that," which is a phrase that can be traced back to a fundamental propositional logic principle called *modus ponens*, which is a Latin phrase meaning "mode that affirms by affirming." This principle may be summarized by this logical rule:

P implies Q and P are both asserted to be true, so therefore Q must be true.

The IFTTT website allows you to create an applet, which is activated by a predefined trigger action sensed by a Home device. The applet will then use a Web service to connect with the target RasPi, which must also be connected to the Internet. You must first register on the IFTTT website in order to create an applet. It is a simple six-step process to create an applet once you have registered and created a free account.

Figure 2-3 shows the initial screen to create an applet. You make this screen appear by clicking on the My Applets icon shown on the main IFTTT introductory screen.

Next, click on the "this" word in the predominant phrase "if +this then that" shown on the screen. The screen shown in Figure 2-4 will next appear. This screen allows you to select a Web service that will successfully interface with the Google Home device. Entering a "G" in the search text box causes all the available Web services with names starting with a G to appear. You will need to click on the Google Assistant icon to select the required Web service. This is precisely the same Web service that enabled the Google Voice Kit to function in the Chapter 1 demonstration. This action completes the first step in the applet creation process.

The second step is to choose a trigger action for the applet. I chose "Say a simple phrase" as the triggering action. You can select this trigger action by clicking on the appropriate block, as shown in Figure 2-5, which automatically appears after completing the first step.

Clicking on the trigger block causes the screen shown in Figure 2-6 to appear. You need to fill in the required fields both to request and response phrases. I chose two very simple and direct phrases, as you can readily see from the figure. Click on Create Trigger to generate the trigger action and to proceed to the next step.

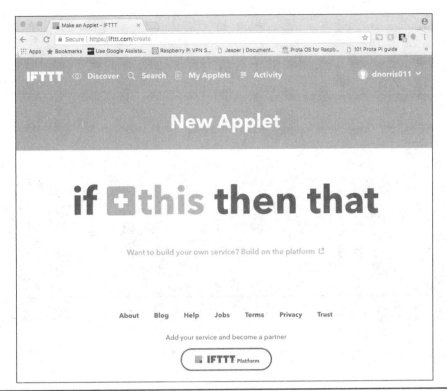

Figure 2-3   Initial screen for applet creation at ifttt.com.

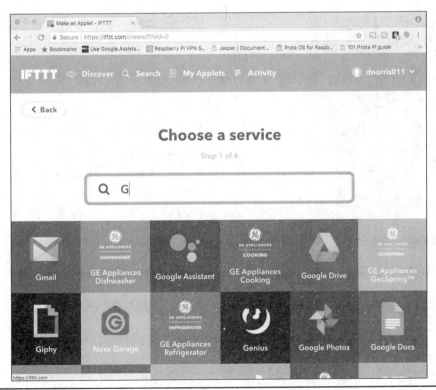

Figure 2-4   Choose a Service selection screen.

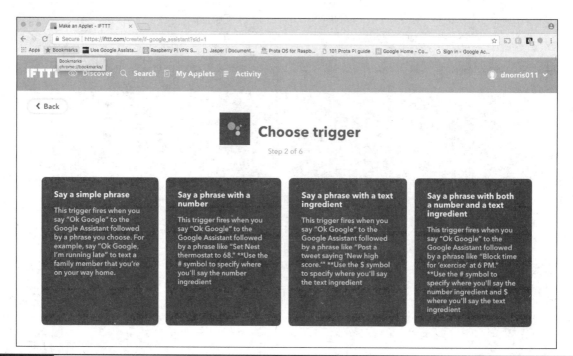

Figure 2-5 Choose Trigger screen.

Figure 2-6 Complete Trigger Fields screen.

You should now see the phrase "if <<icon for selected web service>> then +that" appear, as shown in Figure 2-7. Click on the "that" word to make the screen that allows you to select the Web service used to connect with the RasPi.

Figure 2-8 shows the screen for Choose Action Service that appears, which will allow you to select the appropriate Web action to be taken after the trigger action has completed. Enter "Web" in the search text box, which causes all the available Web services with names starting with Web to appear. In this case, only one service named *Webhooks* will appear. Click on the Webhooks icon, as shown in the figure, to select it. Webhooks is the current name for a Web service previously known as Maker. Figure 2-9 shows the next screen, which is used to start the Web request configuration process. Click on the Make a Web Request box to proceed.

Figure 2-7 Screen to proceed to action selection.

Figure 2-8 Choose Action Service screen.

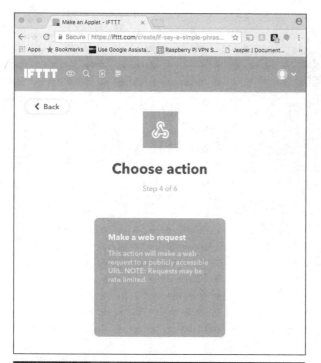

Figure 2-9   Make a Web Request screen.

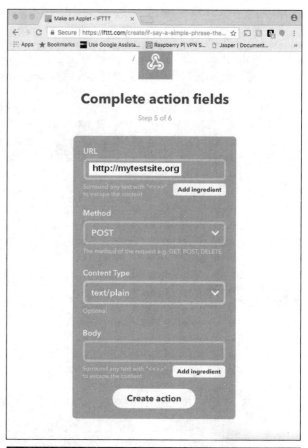

Figure 2-10   Complete Action Fields screen.

Clicking on the Web Request block causes the screen shown in Figure 2-10 to appear. You will first need to fill in the required field for the URL. This URL field raises the interesting question on how to make your RasPi Web server available on the Internet. You probably already realize that the RasPi connects to your home WiFi network using a local IP address. In my case, this was 192.168.1.31. You can easily determine this address by entering the `ifconfig` command in a terminal window. Figure 2-11 shows the result for my situation.

However, the local IP address is not the correct one to enter into the URL text box needed for the applet. You will need your external IP address, which is automatically assigned by your Internet service provider (ISP). You can easily find this by opening a browser on the RasPi and going to the myip.com website. Figure 2-12 shows the result of my search on this website. This is the IP address that must be entered into the applet URL text box, provided

that you are not using a DynDNS type of service, as described below.

You should be aware that there is no guarantee that the external IP address will remain the same because it is dynamically assigned by the ISP each time you connect to your ISP server. This means that you must either edit the appropriate applet(s) each time you run the project or subscribe to a service that provides a constant Web location. The latter approach is the way I chose to handle this issue. I subscribe to the DynDNS service, which provides a URL of my choosing so that all I need to do is type that into the URL text box. This DynDNS-managed URL is constantly updated with my latest ISP-assigned external IP address, automating the entire connection process. You can create any URL you desire that

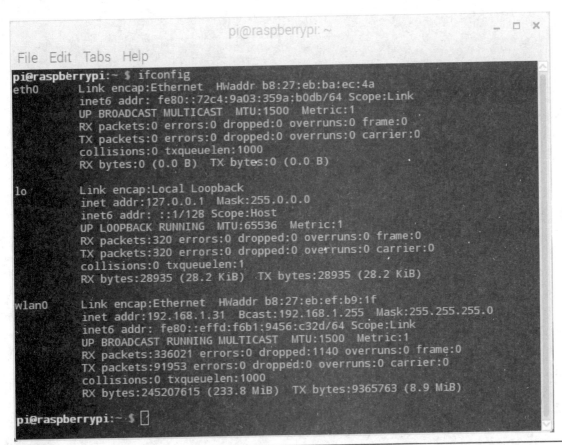

Figure 2-11 Screen results for `ifconfig`.

Figure 2-12 Myip.com screen results.

is not already in use, consistent with normal Internet naming practices. If you examine Figure 2-10, you should see that the URL I entered is http://mytestsite.org (this is not my true URL because I chose to alter it for safety and security purposes). In any case, you need to either enter your currently assigned numeric external IP address or your DynDNS website URL as appropriate.

One further action is required to connect the RasPi on the local network to the external world. You need to enable port forwarding for HTTP packets on your router to the RasPi hosting the Web server. Figure 2-13 shows this port forwarding screen for my Netgear router. Your router will likely have a different screen, but the end result will be the same. Once in place, any external HTTP requests received by

Port Forwarding / Port Triggering

Please select the service type.

- ⦿ Port Forwarding
- ◯ Port Triggering

Service Name

FTP

Server IP Address

192 . 168 . 1 . ___   + Add

	#	Service Name	External Start Port	Internal Start Port	Internal IP address
◯	1	HTTP	80	80	192.168.1.31

✏ Edit Service   ✕ Delete Service   + Add Custom Service   Arrange by Internal IP

Figure 2-13   Port forwarding edit screen.

the router will automatically be forwarded to the designated local IP address.

All that's left in this step is to click on the Create Action button to generate the action. Figure 2-14 shows the final screen in this six-step applet-creation process.

Just click on the Finish button to complete the applet generation. Once you do that, a summary screen similar to Figure 2-15 should appear.

You have the option to click on the Check Now button to have the applet checked for any

Figure 2-14   Review and Finish screen.

Figure 2-15   Summary screen for the newly created applet.

errors or omissions. I have never encountered an error in using this six-step process. You should also observe that the icons for the desired Web services are shown in the Summary screen instead of any actual text identifying those services. This applet is now part of your My Applets collection to be found in your account on the ifttt.com website. Please realize that you do not have to have the ifttt.com site open for this applet to be accessed. Your Google account associated with the Home device has all the links stored in its memory to access both IFTTT and Webhooks. Your trigger phrase will start the process of retrieving all the needed Web links and send the desired Web request to your RasPi Web server. The next section discusses the RasPi Web server and requisite software necessary to respond to the Web request and carry out the desired action(s).

RasPi Web Server Software

This section discusses the required software to process the Web requests received from the Home device. As stated earlier, a number of methods are available to accomplish this task. They range from installing a full-service Web server such as Apache or nginx to using a minimal Python script. I chose the latter because it exposes the fairly simple Web requests that can be easily handled and minimizes the computational load on the RasPi. I will say that if a Web request required storing or retrieving any data from the Home device, I would have chosen the full-service Web server approach. However, this demonstration project is simply transactional, requiring no data to be sent or received.

I had two main purposes in mind for presenting this approach for implementing a Web server on a RasPi. The first was to demonstrate a working Python script that would control the LED as desired. The second,

and perhaps more important, purpose was to present a framework or template that could be used to implement a variety of control schemes extending far beyond the control of a simple LED circuit.

The following Python script handles Web requests to either turn the LED on or off. I have added some explanatory comments both in the script and following it to help clarify what is happening in this program.

```python
# Using RPi.GPIO and Flask for this script
import RPi.GPIO as GPIO
from flask import Flask

app = Flask(__name__)

# This is the default method that is
# invoked without an extension
@app.route("/", methods=['GET', 'POST'])
def index():
    GPIO.setmode(GPIO.BOARD)
    GPIO.setup(12, GPIO.OUT)
    print "Turning the LED on"
    GPIO.output(12, GPIO.HIGH)
    return "LED on"

# This method is invoked when an "/off"
# extension is detected
@app.route("/off", methods=['GET','POST'])
def off():
    GPIO.setmode(GPIO.BOARD)
    GPIO.setup(12, GPIO.OUT)
    print "Turning off the LED"
    GPIO.output(12, GPIO.LOW)
    return "LED off"


if __name__ == "__main__":
    app.run(host='0.0.0.0', port=80,
        debug=True)
```

This Python script is a very simple Web server that will parse out any Web request and run the appropriate Python code that is associated with the request. In this script, there are two Web request formats. The first one does not have an extension or argument and is considered the default form. This default case in the applet created in the preceding section is simply the basic URL. The sample I provided in the applet is http://mytestsite.org without any extensions. This Web request's purpose is to turn on the LED. The statement `@app.route("/", methods=['GET', 'POST'])` detects this default case and allows the code immediately following it to run. The code that follows this statement will be discussed later in this section.

The statement `@app.route("/off", methods=['GET','POST'])` detects a URL with an extension. In this case, the URL with the extension is http://mytestsite.org/off, which is a Web request to turn the LED off. Note that I have not yet shown you how I created an applet to turn off the LED. This is the Web server side with the Python code following the URL parsing statement for actually turning off the LED.

The following Python code snippet turns on the LED and is worth discussing:

```python
def index():
    GPIO.setmode(GPIO.BOARD)
    GPIO.setup(12, GPIO.OUT)
    print "Turning the LED on"
    GPIO.output(12, GPIO.HIGH)
    return "LED on"
```

This is a Python function definition that is executed when a Web request without an extension is detected or parsed. The code uses a Python library called *RPi.GPIO* that permits direct access to the RasPi GPIO pins. Recall that I used the C language to access and manipulate the GPIO pins in Chapter 1. It is also possible to do the same in this script, but it is much easier to simply use Python commands to directly control a GPIO pin. The statement `GPIO.setmode(GPIO.BOARD)` sets up the software to use the physical pin numbers instead of the manufacturer's numbers (BCM mode).

The next statement, `GPIO.setup(12, GPIO.OUT)`, sets physical pin 12 as an output. The next print statement causes an output in the RasPi terminal window to display "Turning the LED on," which provides a positive feedback as to what is happening in the Web server. The next statement, `GPIO.output(12, GPIO.HIGH)`, sets pin 12 to a high state, with 3.3 V appearing on it. The last statement returns "LED on" and returns the string LED onto the website originating the request. It will be ignored in the case of the Webhooks service, but you will see this response if you connect to the RasPi Web server using a browser in lieu of the Google Home device. Remember, the RasPi URL is freely available on the Internet, and you may connect to it using any number of compatible devices, including smartphones, which I demonstrate later in this chapter.

The `off()` function definition is almost identical to the `index()` function definition except that it places pin 12 in an off state or 0 V on the pin. This action will turn off the LED, which is what is expected with the Web request with the off extension added to the URL.

You will need to enter the Python script into the RasPi using a text editor. I strongly recommend using the nano editor to do this. Enter the following command in a terminal window to create the Editor Entry window for a file named *LED_test.py*:

```
sudo nano LED_test.py
```

Now enter the Python script as shown in the listing, paying attention to proper indentation. You can alternatively download this file from this book's website, www.mhprofessional.com/NorrisHomeAutomation.

The program can be run in the Home directory after being entered by executing the following command:

```
sudo python LED_test.py
```

However, there is still one more software task to be done before the project is ready to be tested: you must create an applet to turn off the LED using a voice command. This is easily accomplished using the same six steps previously detailed for creating the applet to turn on the LED. The only significant difference is specifying the action URL. This one must be the same as the URL turning on the LED with the off extension appended. Figure 2-16 shows the new action URL required to turn off the LED.

Failing to include this new applet will simply cause the Home device to respond with the spoken message "Sorry, can't help with that, but I am always learning." This would be a great indication that you forgot to create the applet.

Of course, you must also have the LED circuit set up to view the LED being turned on and off.

Demonstration Circuit

Figure 2-17 is a schematic for the test circuit. It is very similar to Figure 1-16 with the exception that a different GPIO pin is used and the wiringPi library is not used.

Figure 2-18 is a photograph of the physical setup where you can see that I used a T-Cobbler to extend the GPIO pins to a solderless breadboard. No additional physical components are required for this simple demonstration.

Test Run

You will be ready to test the project once the two applets are created, the RasPi software is installed, and the physical circuit is set up. Just execute the RasPi software as shown previously and speak to the Home device with this phrase:

"OK Google, turn on the LED"

If everything has been set up properly, you should be rewarded with seeing the LED turn on. You should also hear the Home device response phrase "The LED is on." In a similar manner, speak the following phrase to turn off the LED:

"OK Google, turn off the LED"

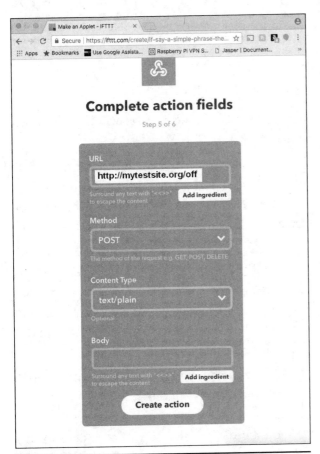

Figure 2-16 Complete Action Fields screen for turning off the LED.

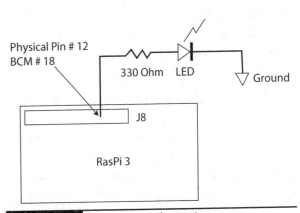

Figure 2-17 Test circuit schematic.

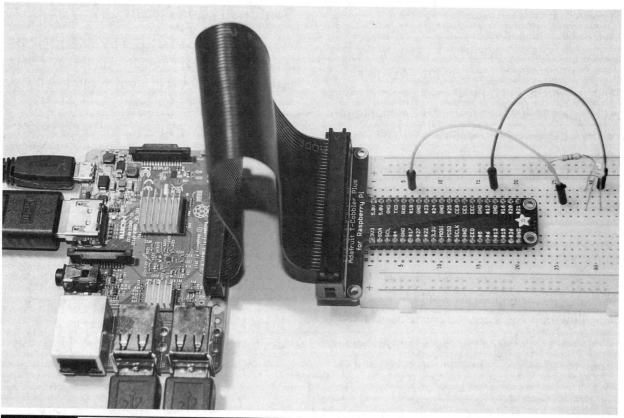

Figure 2-18   Physical test setup.

The LED should turn off after you after spoken the phrase and have heard the response phrase "The LED is off." Please note that it may take a second or two for the action to complete after the phrase has been spoken because of normal delays in activating the various Web services and any latent connectivity issues that might exist between your home network and your ISP provider.

You will also want to test the project using an ordinary Web browser just to confirm what I mentioned earlier that you can turn the LED on or off using a browser. Just remember that there are no Web services involved with this direct action. It is basically a good check on how well the RasPi Web server functions. I entered the turn-on URL in a browser on a computer connected to my local network and saw that the LED turned on. I also saw the "LED on"

message appear in the browser window, as shown in Figure 2-19.

In similar manner, I entered the turn-off URL in the browser and observed the LED turn off. The "LED off" message appeared in the browser screen, as shown in Figure 2-20.

You can also use the RasPi local IP address to test the RasPi server-side software. Just enter the IP address using a browser on another computer attached to your home network. In my case, that would be 192.168.1.31 to turn on the LED and 192.168.1.31/off to turn off the LED.

You can even use a browser on the same RasPi that is concurrently running the Web server. Enter local host in a browser, which should turn on the LED. Enter localhost/off to turn off the LED. Note that you likely forgot to start

Figure 2-19 "LED on" message in a browser window.

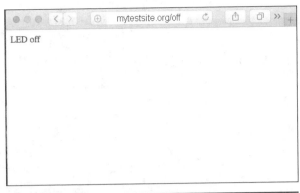

Figure 2-20 "LED off" message in a browser window.

the Web server if you get a message "Connection refused."

Finally, you can also use a browser on a smartphone or a WiFi-connected tablet to control the LED. Just enter the appropriate URL into the smartphone/tablet, and you will be able to turn the LED on and off. The same response messages shown in Figures 2-19 and 2-20 will appear on the smartphone/tablet screen.

Extending Control Functions

At this point, you may be thinking that it is great that a LED can be controlled by a Google Home device, but how is that related to HA? The answer is simple: almost any device that is plugged into the AC mains may be controlled using the same approach taken by the control of a LED. The only difference lies in using different technology to control a device supplying AC power to a given device. The next demonstration will control a common AC-powered lamp using a RasPi GPIO pin. The only difference between this demonstration and the preceding one is that the GPIO pin will be connected to a special AC power control device that has a low-power control circuit already embedded in it.

AC-Powered Lamp Demonstration

The key component in this system design is a Power Switch Tail II power control device. Figure 2-21 shows this device.

This device is essentially a power cord that is controlled by a low-level digital signal that will be directly connected to a RasPi's GPIO pin. The Power Switch Tail II, which I will now refer to as the *PST2*, uses an optically isolated digital input to control a power relay capable of handling up to 15 A at 120 VAC. The optical isolation eliminates any safety concerns about dealing with mains-type power with the RasPi

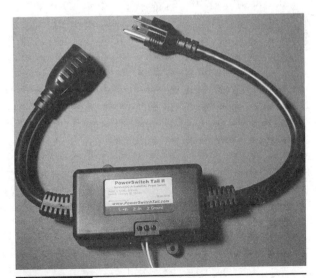

Figure 2-21 Power Switch Tail II.

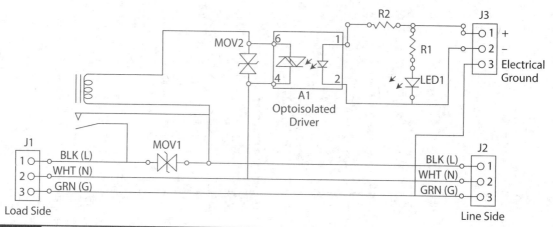

Figure 2-22   PST2 schematic.

board. The PST2 is also ruggedly constructed and very well insulated, making it extremely safe to use in a home environment. This power control device can also handle loads up to 1.5 kilowatts (kW), which is well beyond anything I will use in this demonstration.

The PST2 schematic is shown in Figure 2-22, which points out the robust and safe design that makes up the PST2. I highly recommend that you purchase an appropriate number of PST2s to control any AC power loads you may be considering for both safety and convenience sake. Building your own power controller is

really not a good idea. The PST2 has already passed all appropriate UL and other safety certifications and is ready to use.

Figure 2-23 is a simple schematic showing how the RasPi pin 7 connects to a PST2. Nothing is required for the connection other than a pair of wires carrying the GPIO signal and ground. In this case, I chose physical pin 7 (BCM pin 4) for the control signal. GPIO 4 (pin 7) connects to the PST2 (+ in) terminal, and the RasPi GND (pin 6) connects to the PST2 (– in) terminal.

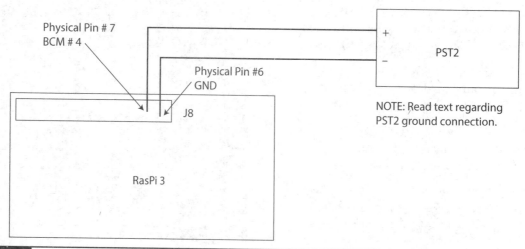

Figure 2-23   RasPi-to-PST2 connection schematic.

CAUTION: Do not connect the PST2 ground terminal to the RasPi ground. Simply leave it unconnected. It is not required nor needed, and it could possibly be an entry point for mains power if there were some odd and strange failure on the PST2 load side.

All that is required now is to describe the two applets required to control the AC lamp and the corresponding modifications to the RasPi Python Web server.

AC Lamp Applets

Two applets are required to have the Home device control the AC lamp. The same six-step applet creation that we used for LED control is used for AC lamp control. The only differences are detailed in Table 2-1.

TABLE 2-1 AC Lamp Control Applet Parameters

Applet item	Turn-on applet	Turn-off applet
Request phrase	Turn on AC lamp	Turn off AC lamp
Response phrase	The AC lamp is on	The AC lamp is off
URL	http://mytestsite.org/AClampon	http://mytestsite.org/AClampoff

Python Web Server Modifications

Two additional code snippets must be added to the existing Python script to handle the new Web requests that are received for AC lamp control. The complete modified script is listed below. I also renamed the script *extended_test.py* to reflect the newly added functionalities concerning the AC lamp.

```
# Using RPi.GPIO and Flask for this script
import RPi.GPIO as GPIO
from flask import Flask

app = Flask(__name__)

# This is the default method that is invoked without an extension
@app.route("/", methods=['GET', 'POST'])
def index():
    GPIO.setmode(GPIO.BOARD)
    GPIO.setup(12, GPIO.OUT)
    print "Turning the LED on"
    GPIO.output(12, GPIO.HIGH)
    return "LED on"

# This method is invoked when an "/off" extension is detected
@app.route("/off", methods=['GET','POST'])
def off():
    GPIO.setmode(GPIO.BOARD)
    GPIO.setup(12, GPIO.OUT)
    print "Turning off the LED"
    GPIO.output(12, GPIO.LOW)
    return "LED off"

# This method is invoked when an "/AClampon" extension is detected
@app.route("/AClampon", methods=['GET','POST'])
def AClampon():
```

```
GPIO.setmode(GPIO.BOARD)
GPIO.setup(7, GPIO.OUT)
print "Turning on the AC lamp"
GPIO.output(7, GPIO.HIGH)
return "AC lamp on"

# This method is invoked when an "/AClampoff" extension is detected
@app.route("/AClampoff", methods=['GET','POST'])
def AClampoff():
    GPIO.setmode(GPIO.BOARD)
    GPIO.setup(7, GPIO.OUT)
    print "Turning off the AC lamp"
    GPIO.output(7, GPIO.LOW)
    return "AC lamp off"

if __name__ == "__main__":
    app.run(host='0.0.0.0', port=80, debug=True)
```

Test Run

You will be ready to test this new project once the two additional applets are created, the modified RasPi software is installed, and the physical circuit is set up. Just execute the RasPi software with

```
sudo python extended_test.py
```

Ensure that the PST2 is plugged into the AC mains and an ordinary table lamp is connected to the PST2. Also ensure that the lamp has been turned on. It will not light because the PST2 has not connected it to the AC mains. Now speak to the Home device with this phrase:

"OK Google, turn on AC lamp"

If everything has been set up properly, you should be rewarded with seeing the lamp turn on and hearing the response phrase "The AC lamp is on." In a similar manner, speak the following phrase to turn off the lamp:

"OK Google, turn off AC lamp"

The lamp should turn off after you have spoken the phrase along with hearing the response phrase "The AC lamp is off." Figure 2-24 is a terminal window screen shot showing the print

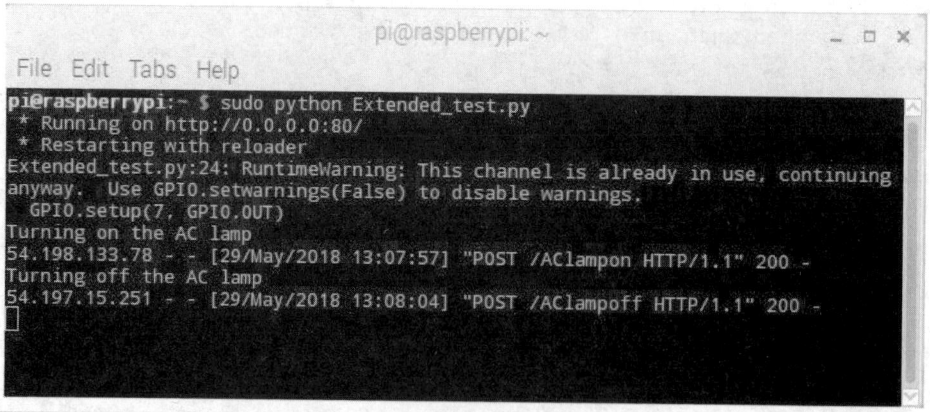

Figure 2-24 AC lamp terminal window control print statements.

statements associated with the Web requests for the AC lamp.

You may have already realized by now that there is a significant disadvantage to using the PST2 in the manner described in this project. The issue is that a wire-pair cable must be connected between the RasPi and the PST2 in order to control the power device. This may not be a problem if you only have one RasPi dedicated to one powered device. However, it becomes problematic if you want to control multiple loads that are situated relatively far apart or even in separate rooms. Such a situation requires running long lengths of wire cables to different PST2s. This approach is not practical and would be unsightly and unacceptable for most HA installations. A much better, very inexpensive wireless solution is presented in the next section.

Transmitter

Receiver

Figure 2-25 A 433-MHz transmitter/receiver pair.

PST2 Wireless Control

It is a simple process to extend the range of control between a RasPi and a PST2 power device using wireless technology. I chose to use a simple radio-frequency (RF) transmitter/receiver pair that operates in the unlicensed 433-MHz band. Figure 2-25 shows an inexpensive wireless 433-MHz RF module receiver and transmitter remote control pair with built-in learning codes for four-channel operation suitable for use with a RasPi.

There is a coiled wire antenna that is 52 centimeters (cm) long for both the transmitter and the receiver. These antennas were specifically cut to this length for tuned operation at 433 MHz. The key to using this design is that only an on/off signal needs to be sent between the RasPi and the PST2. This solution will not work if any complex data are required to be sent or a response signal is sent. However, I will be discussing a much more elegant wireless solution

in Chapter 10, where more sophisticated control signals are required to implement a wireless HA system.

Wireless Communications Setup

The transmitter will generate an encoded RF signal whenever a 5-V signal is applied to one of its four data input pins. Figure 2-26 is a connection block diagram for the complete communications link showing both the transmitter and receiver connected to the RasPi and PST2, respectively.

Please note that I used a 5-V wall-wart power supply to provide power to the receiver. The receiver was installed on a small solderless breadboard, which, in turn, was mounted on the PST2 using double-backed tape. The physical setup is shown in Figure 2-27.

It would be a simple matter to mount the receiver in a plastic box for aesthetics and to prevent any inadvertent damage to the exposed

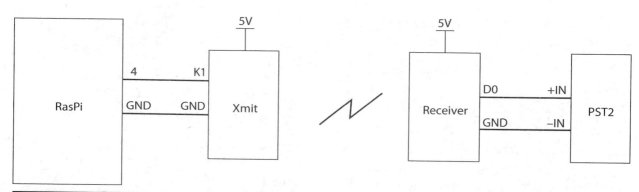

Figure 2-26   Wireless connection block diagram.

Figure 2-27   PST2 with wireless receiver.

components. Note that no dangerous voltages are exposed in the breadboard version.

You should also realize that this system design will support four separate remote PST2 devices, where each device has its own dedicated RF channel. These channels will not interfere with each other because a simple digital encoding scheme is used between the transmitter and each receiver. All you need to do is copy my single-channel implementation to any of the other three channels. The only additional software needed would be the applets and RasPi Web server changes required to support the additional devices control by the extra channels.

My only caution is that interference might be present if you decide to use another four-channel transmitter/receiver pair in the same vicinity as this one. Inexpensive transmitter/receiver pairs such as the one I used usually do not incorporate interference safeguards from similar units.

Figure 2-28 shows how power and input signals are applied to the transmitter. I used the K1 input, which shows a pushbutton that when pressed will connect that input to ground. This action means that I had to change the RasPi output on GPIO pin 4 (BCM mode) from normally LOW to normally HIGH. This was a fairly simple change and is shown in the following revised *extended_test.py* code listing,

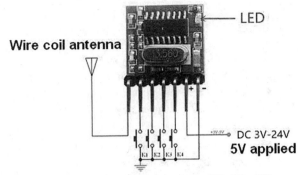

Note: Control signal applied to K1

Figure 2-28 Wireless link power and signal connections.

which I renamed *remote_test.py* to indicate its new purpose:

```python
import time
# Using RPi.GPIO and Flask for this script
import RPi.GPIO as GPIO
from flask import Flask

# Required to set GPIO physical pin 7 normally HIGH
GPIO.setmode(GPIO.BOARD)
GPIO.setup(7, GPIO.OUT)
GPIO.output(7, GPIO.HIGH)


app = Flask(__name__)

# This is the default method that is invoked without an extension
@app.route("/", methods=['GET', 'POST'])
def index():
    GPIO.setmode(GPIO.BOARD)
    GPIO.setup(12, GPIO.OUT)
    print "Turning the LED on"
    GPIO.output(12, GPIO.HIGH)
    return "LED on"

# This method is invoked when an "/off" extension is detected
@app.route("/off", methods=['GET','POST'])
def off():
    GPIO.setmode(GPIO.BOARD)
    GPIO.setup(12, GPIO.OUT)
    print "Turning off the LED"
    GPIO.output(12, GPIO.LOW)
    return "LED off"

# This method is invoked when an "/AClampon" extension is detected
```

```
@app.route("/AClampon", methods=['GET','POST'])
def AClampon():
    GPIO.setmode(GPIO.BOARD)
    GPIO.setup(7, GPIO.OUT)
    print "Turning on the AC lamp"
    # Next three statements send a 0.25 second "LOW" pulse
    GPIO.output(7, GPIO.LOW)
    time.sleep(0.25)
    GPIO.output(7, GPIO.HIGH)
    return "AC lamp on"

# This method is invoked when an "/AClampoff" extension is detected
@app.route("/AClampoff", methods=['GET','POST'])
def AClampoff():
    GPIO.setmode(GPIO.BOARD)
    GPIO.setup(7, GPIO.OUT)
    print "Turning off the AC lamp"
    # Next three statements send a 0.25 second "LOW" pulse
    GPIO.output(7, GPIO.LOW)
    time.sleep(0.25)
    GPIO.output(7, GPIO.HIGH)
    return "AC lamp off"

if __name__ == "__main__":
    app.run(host='0.0.0.0', port=80, debug=True)
```

The modification details to the *extended_test.py* script include

- Making the default output to the 433-MHz receiver a HIGH state

- Changing the AClampon and AClampoff script portions to emit a 0.25-second pulse instead of a continuous level either HIGH or LOW

You may have noticed that the AClampon and AClampoff portions are now nearly identical except for the response messages. This is because the RF receiver is set to be in a latching or toggle mode, which means the state changes form HIGH to LOW or LOW to HIGH for every received pulse. I could have used only a single applet to accommodate this situation, but that would have meant using a single phrase such as

"OK Google, toggle the AC lamp" to switch the lamp mode. I chose not to use such a phrase because many nontechnical users would have no idea what was being requested, and the user would already need to know the current lamp's state. The lesson for this situation is sometimes that you have to be more expansive in your coding to simplify the user's experience.

Figure 2-29 is a close-up photograph of the four-channel receiver in which you can clearly see where you will need to solder the wire coil antenna. The two connection points going to the PST2 are also clearly indicated. The receiver's Learning button is shown in Figure 2-30.

The receiver must be set up in the toggle or latching mode in order for the system to work properly. The following steps set the receiver to a particular mode:

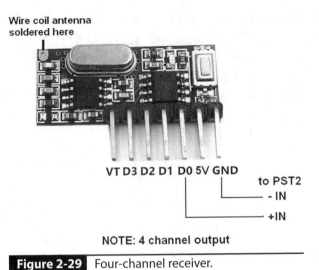

Wire coil antenna
soldered here

VT D3 D2 D1 D0 5V GND

to PST2
- IN
+IN

NOTE: 4 channel output

Figure 2-29   Four-channel receiver.

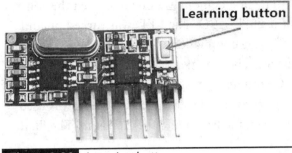

Learning button

Figure 2-30   Learning button.

1. Press the Learning button once to set the momentary mode. This means that the channel output will remain LOW for as long as a signal is being received.

2. Press the Learning button twice to set the toggle mode. This means that the channel output will change state, that is, from HIGH to LOW or LOW to HIGH, and remain in the new state.

3. Press the Learning button three times to set the interlocked mode. This means that the selected channel will be active, and all other channels will become inactive.

Test Run

I connected the AC lamp to the PST2 in exactly the same way as I did in the preceding demonstration. I next plugged in the 5-V wall-wart power supply for the receiver to power it on. I next pressed the Learning button twice to set it to the latching or toggle mode. You will only have to set the mode one time. The receiver "remembers" the setting even after the 5-V power has been turned off and on.

The AC lamp turned on and stayed on after the trigger phrase was spoken to the Home device. Likewise, the lamp turned off when the off-action trigger phrase was spoken. This last demonstration concludes this chapter.

Summary

This chapter started with a detailed discussion of how the Google Home Personal Voice Assistant works. I also described how to set up the Home device in its normal configuration.

The next section described how to logically connect the Home device to a RasPi. The first step in the connection process was to use the ifttt.com website to create a Web applet. This applet converts a voice command into an actionable Web request. This Web request is then sent over the Internet to a Web server hosted on the HA RasPi. A very detailed discussion was provided concerning the six-step process used to create the Web applets.

The RasPi Web server software was next described. A Python script was listed that handles the incoming Web requests and creates appropriate responses using the GPIO subsystem.

The first demonstration concerned the control of a single LED. The LED was turned on and off using commands spoken into the Home device. The next demonstration controlled an AC lamp using a similar voice command. A power control device called the PST2 was used between the RasPi and the AC lamp. The final demonstration was an extension of the second in which PST2 control was remotely extended using a four-channel transmitter/receiver pair. This arrangement allowed for great flexibility in the control of mains-powered devices in an HA environment.

## CHAPTER 3

Raspberry Pi Implements a Google Voice Assistant

CHAPTER 1 SHOWED YOU HOW to connect the inexpensive Google Voice Kit to a RasPi and have the RasPi respond to spoken voice commands. Chapter 2 showed you how to interconnect a Google Home device to a RasPi and again have the RasPi respond to spoken voice commands detected by the Home device. In this chapter, I will show you how to use the RasPi all by itself to emulate a Google Voice Assistant and respond directly to spoken voice commands. This approach will be the least costly of all the Google Voice Assistant implementations but will likely require the most effort on your part to duplicate the demonstrations.

I would also like to extend my appreciation to several Google Home–type device developers and especially to Keval Patel for his great July 2017 blog, "Turn Your Raspberry Pi into a Homemade Google Home." I used a good deal of information from this blog, although I had to make several modifications to make an operable system. This is not a criticism of Keval but instead is a consequence of using open-source software, which is dynamic and constantly changing.

Audio Setup

The initial steps in building your own Google Home assistant involve setting up the audio components supporting the project. The RasPi audio output should be forced to use the 3.5-mm

Parts List

Item	Model	Quantity	Source
RasPi 3	B or B+	1	adafruit.com amazon.com mcmelectronics.com
USB microphone	Commodity	1	amazon.com
USB speaker	Commodity	1	amazon.com
Power Switch Tail	II	1	amazon.com
AC table lamp	Commodity	1	Home improvement store
Hobby-grade servo	HiTEC HS-311	1	amazon.com
6-V DC power supply	Commodity	1	adafruit.com

AV jack located on the board. This is done by entering the following command in a terminal window:

```
sudo raspi-config
```

Select Advanced Options, and then click on the Audio option. Next, click on

```
2. Force audio output 3.5 mm jack
```

Any audio signals generated by the RasPi will now be directed to the onboard AV jack. You should now ensure that the amplified USB speaker(s) is(are) plugged into the RasPi. Next, enter the following command into the terminal window:

```
speaker-test -t wav
```

You should start hearing "front left" coming from the speaker(s) and also see the status messages shown in Figure 3-1.

The next portion of the setup concerns the USB microphone. I used a large, semiprofessional USB microphone for this project, which is shown in Figure 3-2. I used this simply because I had already purchased it for use in an earlier AV project. You can choose to purchase a similar device or a much more inexpensive model, such as the one shown in Figure 3-3.

The inexpensive model also should be adequate for this project because all it has to do is detect spoken voice commands. The more expensive model had to adequately record both music and voice signals.

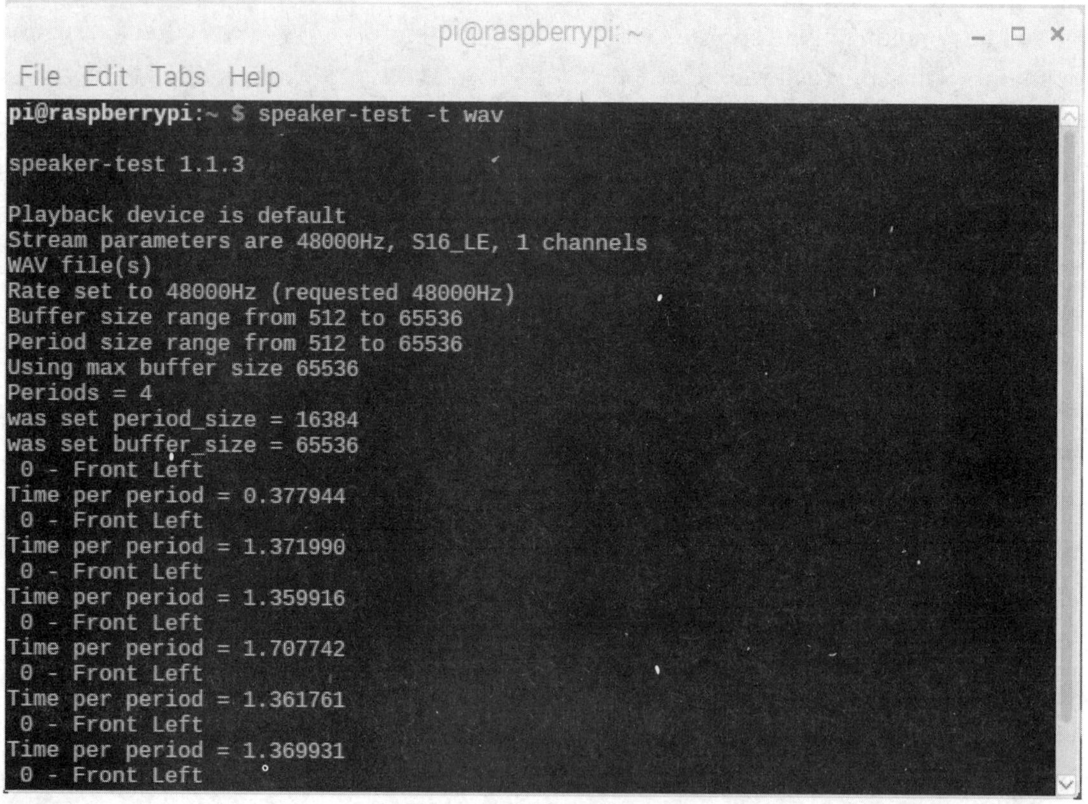

Figure 3-1   Speaker test screen display.

Figure 3-2 USB microphone.

Figure 3-3 Inexpensive USB microphone.

In either case, just enter the following command to record a brief audio clip to test the USB microphone:

```
arecord --format=S16_LE --duration=10
--rate=16000 --file-type=raw test.raw
```

Then speak directly into the microphone for approximately 10 seconds. It really does not

matter what you say because it will be recorded in an audio file for an immediate payback. The playback is initiated by the following command:

```
aplay --format=S16_LE --rate=16000
    test.raw
```

You should start hearing the recording that you just made through the attached speaker(s). Recheck the RasPi configuration settings and the component connections if you do not hear the recording.

Installing and Configuring the Python 3 Environment

The Google Assistant software suite requires that a Python 3 environment be installed and configured on the RasPi. The initial step is to ensure that all the current software is updated. Enter the following command in a terminal window:

```
sudo apt-get update
```

The next command sequence will install and set up a Python 3 environment:

```
sudo apt-get install python3-dev
    python3-venv
python3 -m venv env
env/bin/python -m pip install --upgrade
    pip setuptools
```

Certain required Python dependencies were also installed using this command sequence.

You will now need to activate the Python environment by entering this next command:

```
source env/bin/activate
```

You should notice that a virtual Python environment is now in place because the terminal prompt is prepended with (env).

The Google Assistant Software Development Kit (SDK) created for the RasPi now needs to be downloaded and installed. Enter this command to do that:

```
python -m pip install --upgrade
    google-assistant-library
```

The Google Assistant software should now be installed on the RasPi. It is now time to focus on the Google cloud portion of the system.

Installing and Enabling the Google Cloud Project

There are number of steps in this installation process, some of which are duplicates of steps taken in the Cloud installation process shown in Chapter 1. I will go through all the steps to ensure that the project is set up properly.

The first step is log onto your Google account. If don't have an account, you could not have completed the Chapter 1 project, and hence, all the following steps are required. Go and create an account at https://accounts.google.com/SignUp?hl=en; otherwise, log into your existing account.

Next, go to the Google Cloud Platform website at https://console.cloud.google.com/getting-started and create a new project. This is easily accomplished by clicking on the drop-down arrow in the Chapter textbox located next to the Google Cloud Platform title. At least one project should be displayed in the listing, reflecting the Chapter 1 demonstration. Figure 3-4 shows the screen after I clicked on the drop-down arrow.

Click on the New Project button located at the upper right-hand corner of the screen to start a new project. You first have to provide a project

Figure 3-4   New project screen.

name. You may choose any name that makes sense to you. I chose the name *My Google Home* to reflect the project being built. It will take a few seconds for the Google Cloud Platform website to create the project. Once created, you will be returned to the preceding screen.

Once again, click on the drop-down arrow in the Projects textbox, and select the newly created project. You will be shown a configuration screen, a portion of which is shown in Figure 3-5. You will need to click on Enable APIs and get credentials such as keys in the Getting Started section.

You will first need to enable the Google Assistant API. You can do this by clicking on the project Dashboard icon and then clicking on the Enable APIs and Services button. You should next enter `Google Assistant` in the search

textbox to bring up the appropriate service. Click on the Google Assistant API box once it appears, and finally click on the Enable button. This last series of actions will link the Google Assistant services to your Google account, thus enabling the voice-recognition function as well as all the other Google Home functions.

Next, click on Credentials in the column labeled APIs & Services, which should have appeared on the browser after the Google Assistant services were enabled. This action will take you to another screen where you actually create the project credentials by clicking on the drop-down arrow in the Create Credentials screen, as shown in Figure 3-6.

Select OAuth Client ID to proceed, as shown in Figure 3-7.

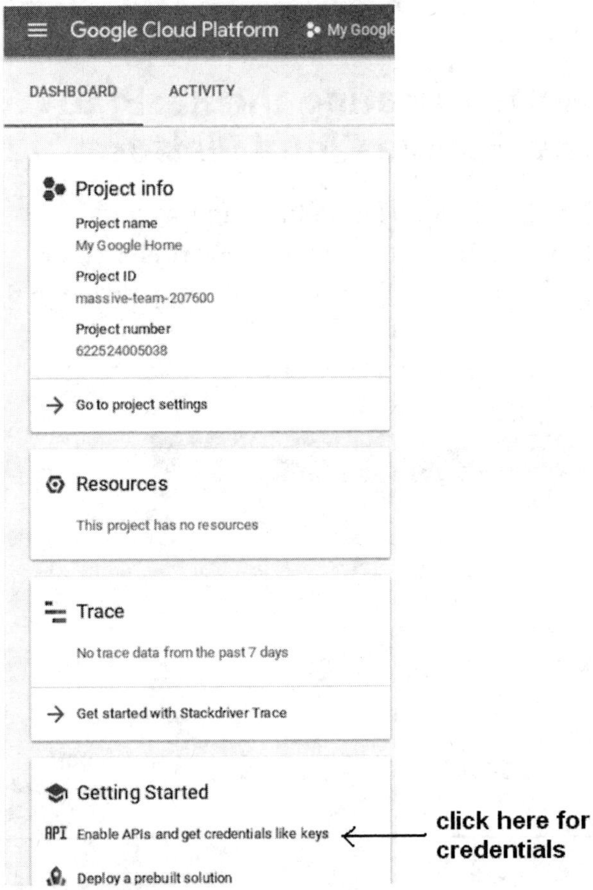

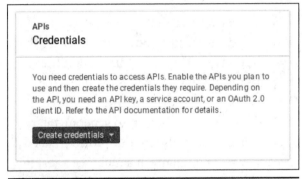

Figure 3-6   Create Credentials screen.

Figure 3-7   Credentials selection screen.

Figure 3-5   API setup screen.

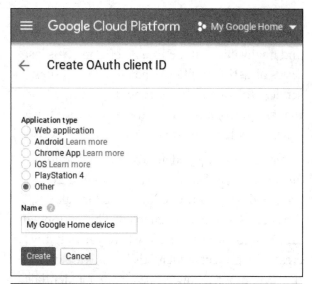

Figure 3-8   Create OAuth Client ID screen.

The Create OAuth Client ID screen appears, as shown in Figure 3-8.

Enter your project name as I did for my project, and ensure that you select Other for the Application type. This is very important because the project will not build correctly unless the Other type has been selected. You now need to click on the Create button to generate the appropriate credentials for your project. This will take a few seconds, and you should see a results screen similar to the one shown in Figure 3-9.

You will now need to download the resulting JSON file to the RasPi. Notice the annotation in the figure that points out the download icon. Just expand the screen if you do not see the icon. The JSON file will be downloaded to the RasPi after you click on the icon. It typically will be stored in the Downloads directory. You will next need to rename it and move it into the Home directory. The default JSON name is incredibly complex, and you should not try to copy it manually. Instead, enter the following commands:

```
cd Downloads
mv client_secret_ <press the tab key for
    auto completion> ~/assistant.json
```

These commands will move the new credentials file from the Downloads directory to the Home directory and rename it `assistant.json`, a much easier file name to handle.

Authenticating the RasPi to the Google Cloud Platform

You first need to install the authorization tool. Ensure that you are in the virtual Python environment. I will now include the prepend to

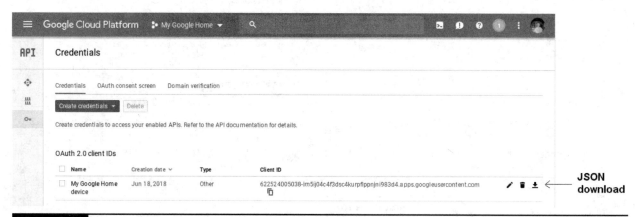

Figure 3-9   Generated credentials.

indicate this situation. Do not enter it as part of the next command:

```
(env) python -m pip install
   --upgrade google-auth-oauthlib[tool]
```

Next, execute the tool using the following command:

```
(env) google-oauthlib-tool
   --client-secrets "PATH_TO_YOUR_JSON_
   FILE" --scope https://googleapis.com/
   auth/assistant-sdk-prototype
   --save --headless
```

In my case, I substituted `~/assistant.json` for `"PATH_TO_YOUR_JSON_FILE"`. You should do the same if you followed my previous download instructions.

The result of this last command is an automatically generated authentication URL, as shown in Figure 3-10. You should copy and paste the authentication URL into a browser. Do not try to copy the complex URL manually.

The authentication URL will create a complex authorization code that you should copy and paste back into the terminal window at the prompt Enter the Authorization Code. The authentication tool, which is still running on the RasPi, will now automatically place the JSON credentials file into the appropriate directory that the Google Assistant requires to grant permission for normal functions.

LED Operations Indicator

A LED will indicate when an active conversation is happening with the Google Assistant service. The LED is connected to GPIO pin 25 (BCM mode), as shown in the Fritzing diagram (Figure 3-11).

The LED indicator is activated by means of a Google Assistant callback function. Callback functions are events that are triggered by specific actions. For instance, the `EventType.ON_CONVERSATION_TURN_STARTED` callback will be triggered when the conversation with the Google Assistant begins. Likewise, the `EventType.ON_CONVERSATION_TURN_FINISHED` callback will be triggered whenever the conversation is terminated. LED indicator activation is tied directly to each of these callbacks, as you will shortly see in the Python script listing.

It is important to realize that LED activation or inactivation is triggered by a state change happening with the Google Assistant software. It is not directly related to actual data but instead to the state of data. State of data is called *metadata* because it reflects some property of data, not the actual value or content of those data. Using metadata for home automation (HA) control is problematic at best because of severe limitations on what can be achieved using strictly metadata. Some readers may recall a mains power device called the Clapper

```
(env) pi@raspberrypi:~ $ google-oauthlib-tool  --client-secrets ./assistant1.jso
n  --scope https://www.googleapis.com/auth/assistant-sdk-prototype --save --headl
ess
Please visit this URL to authorize this application: https://accounts.google.com
/o/oauth2/auth?response_type=code&client_id=622524005038-bc4grklqofi5uk8ebu19evq
o14e5g3v7.apps.googleusercontent.com&redirect_uri=urn%3Aietf%3Awg%3Aoauth%3A2.0%
3Aoob&scope=https%3A%2F%2Fwww.googleapis.com%2Fauth%2Fassistant-sdk-prototype&st
ate=CWY6wBuh9yCs3jCylyvbsVzpCyOFbt&prompt=consent&access_type=offline
Enter the authorization code: 
```

Figure 3-10 Authentication URL.

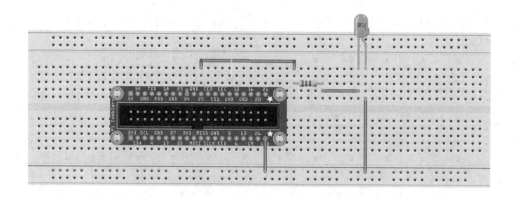

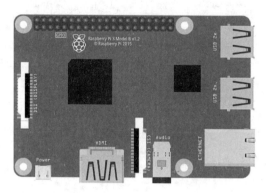

Figure 3-11 LED indicator Fritzing diagram.

that was widely advertised on TV many years ago. The user would clap his or her hands to turn on a device such as an old-fashion TV and clap again to turn it off. No actual data were received by the clapper device other than a noise impulse created by the hand clap. The presence or absence of impulse noise was effectively the metadata for the clapper. I believe that it worked, but eventually it was superseded by much more flexible remote controls that actually transmitted real data in lieu of metadata.

Python Script

The RasPi requires a program to both authenticate and initialize the Google Assistant

software. This is accomplished by the following Python script. However, the RPi.GPIO library must first be installed to allow for programmatic control of the GPIO pins. This installation will happen after you enter the following command:

```
pip install RPi.GPIO
```

Once you have installed RPi.GPIO, you will be ready to enter the Python script. Either enter the script into the nano editor or simply download it from this book's companion website, www.mhprofessional.com/NorrisHomeAutomation. The script is named *main.py* and must be stored in the Home directory.

```python
#!/usr/bin/env python

from __future__ import print_function

import argparse
import os.path
import json

import google.oauth2.credentials
import RPi.GPIO as GPIO
from google.assistant.library import Assistant
from google.assistant.library.event import EventType
from google.assistant.library.file_helpers import existing_file

GPIO.setmode(GPIO.BCM)
GPIO.setup(23, GPIO.OUT)

def process_event(event):
    """Pretty prints events.
Prints all events that occur with two spaces between each new conversation and a single
space between turns of a conversation.
    Args:
        Event(event.Event): The current event to process.
    """
    if event.type == EventType.ON_CONVERSATION_TURN_STARTED:
        print()
        GPIO.output(25,True)

    print(event)

if (event.type == EventType.ON_CONVERSATION_TURN_FINISHED and event.args and not
event.args['with_follow_on_turn']):
        print()
        GPIO.output(25,False)

def main():
    parser = argparse.ArgumentParser(
        formatter_class=argparse.RawTextHelpFormatter)
    parser.add_argument('--credentials', type=existing_file,
                        metavar='OAUTH2_CREDENTIALS_FILE',
                        default=os.path.join(
                            os.path.expanduser<'/home/pi/.config'),
                            'google-oauthlib-tool',
                            'credentials.json'
                        ),
                        help='Path to store and read OAuth2 credentials')
    args = parser.parse_args()
    with open(args.credentials, 'r') as f:
        credentials = google.oauth2.credentials.Credentials<token=None,
                                                  **json.load(f))
```

```
with Assistant(credentials,"My Google Home device") as assistant:
    for event in assistant.start():
        process_event(event)


if __name__ == '__main__':
    main()
```

TIP: The above listing contains very long sets of instructions that display on multiple lines in the listing. The correct script requires that the long instructions sets be entered without line breaks and in accordance with Python's indention format. I highly recommend that you download the script from this book's website unless you are very comfortable in writing Python scripts.

You will now need to create a shell script that will initialize and run the Google Assistant software. Enter the following command to start the nano editor with script named *google-assistant-init.sh*:

```
sudo nano google-assistant-init.sh
```

Next, enter the following into the editor:

```
#!/bin/sh
/home/pi/env/bin/python3 -u
    /home/pi/main.py
```

Save the file and exit the editor. Now grant executable permissions to the shell script by entering this command:

```
sudo chmod +x google-assistant-init.sh
```

You are now finally ready to test the system. Ensure that everything is connected, including the USB microphone and speaker(s).

Test Run

Enter the following command to start the Google Assistant on the RasPi:

```
bash google-assistant-init.sh
```

Next, speak this phrase clearly into the microphone: "OK Google, what time is it?" You should be delighted to hear the words "The time is five forty-nine." Of course, your time will be different, but the system should respond. In addition, you should have observed the LED being lit when you started the conversation. It should have gone off after the system responded with the time, indicating that the conversation had ended.

Conversations may be brief, as in the case of requesting the time, or can be extended if the response is lengthy. Look at Figure 3-12, where the weather report was quite extensive. This conversation lasted over 10 seconds.

At this point, I have presented everything you need to build your own Google Home device using only a RasPi with two USB peripherals. This project is a good emulation of the commercial Google Home unit, but it is still quite limited in its usefulness for HA applications. The next section discusses how to significantly increase the device's flexibility and utility, especially for use in HA systems.

Extending the RasPi Google Home Device

The device as configured at this point has a very limited use in an HA system. It could function in a "clapper-like" mode, as I explained earlier. That restricted operational mode was due to the device only responding to state changes

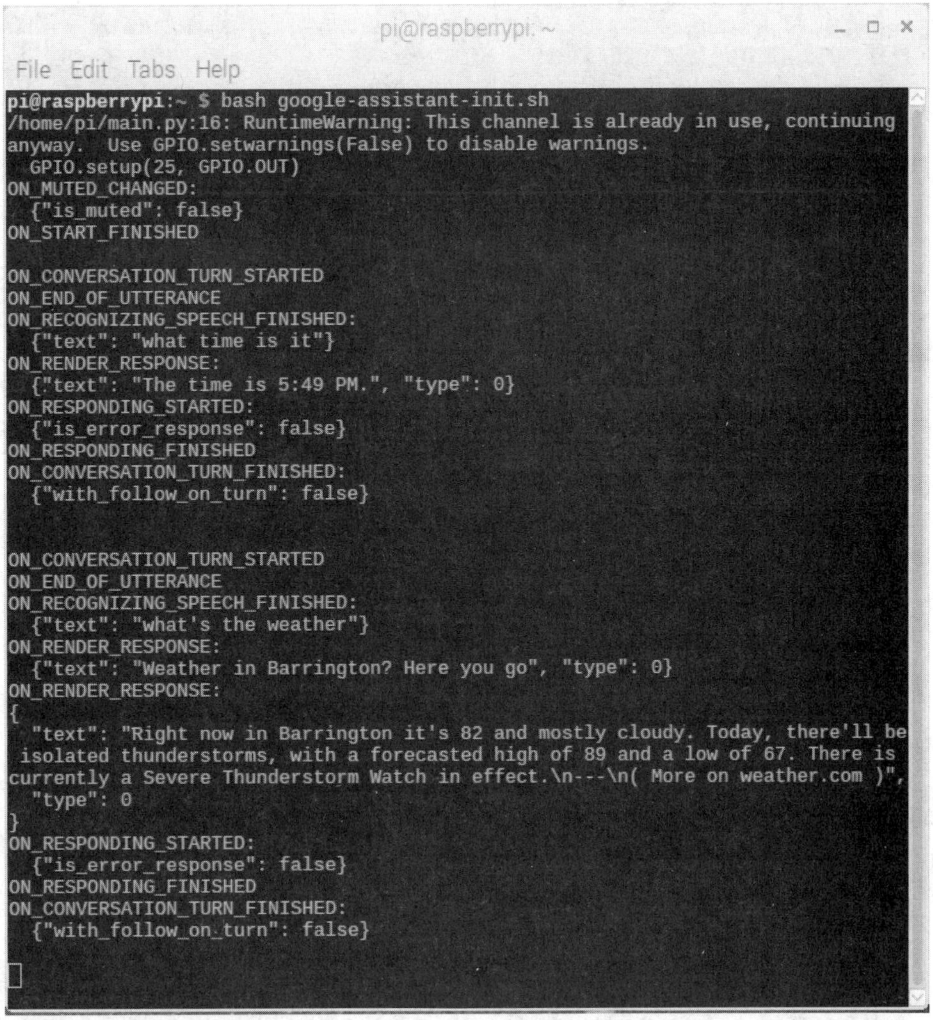

```
pi@raspberrypi: ~                                    _ □ ✕

File  Edit  Tabs  Help

pi@raspberrypi:~ $ bash google-assistant-init.sh
/home/pi/main.py:16: RuntimeWarning: This channel is already in use, continuing
anyway.  Use GPIO.setwarnings(False) to disable warnings.
  GPIO.setup(25, GPIO.OUT)
ON_MUTED_CHANGED:
  {"is_muted": false}
ON_START_FINISHED

ON_CONVERSATION_TURN_STARTED
ON_END_OF_UTTERANCE
ON_RECOGNIZING_SPEECH_FINISHED:
  {"text": "what time is it"}
ON_RENDER_RESPONSE:
  {"text": "The time is 5:49 PM.", "type": 0}
ON_RESPONDING_STARTED:
  {"is_error_response": false}
ON_RESPONDING_FINISHED
ON_CONVERSATION_TURN_FINISHED:
  {"with_follow_on_turn": false}


ON_CONVERSATION_TURN_STARTED
ON_END_OF_UTTERANCE
ON_RECOGNIZING_SPEECH_FINISHED:
  {"text": "what's the weather"}
ON_RENDER_RESPONSE:
  {"text": "Weather in Barrington? Here you go", "type": 0}
ON_RENDER_RESPONSE:
{
  "text": "Right now in Barrington it's 82 and mostly cloudy. Today, there'll be
  isolated thunderstorms, with a forecasted high of 89 and a low of 67. There is
currently a Severe Thunderstorm Watch in effect.\n---\n( More on weather.com )",
  "type": 0
}
ON_RESPONDING_STARTED:
  {"is_error_response": false}
ON_RESPONDING_FINISHED
ON_CONVERSATION_TURN_FINISHED:
  {"with_follow_on_turn": false}
```

Figure 3-12 Conversation log.

conveyed by metadata. Fortunately, the Google Assistant software provides the capability to parse the actual data present in a conversation. This capability will allow for a great increase in the utility of the device, especially for HA applications. I have created duplicates of two demonstrations that were presented in Chapter 2. The first one deals with control of a LED and the second one with control of an AC lamp. The second demonstration also uses a Power Switch Tail II (PST2), which I introduced in Chapter 2.

Test Setup

Figure 3-13 is a Fritzing diagram showing how to set up both the LED and the PST2 peripherals with the RasPi. You should note that the original LED connected to GPIO pin 25 is also shown in the diagram. Table 3-1 details all the connections between the T-Cobbler and the test peripherals.

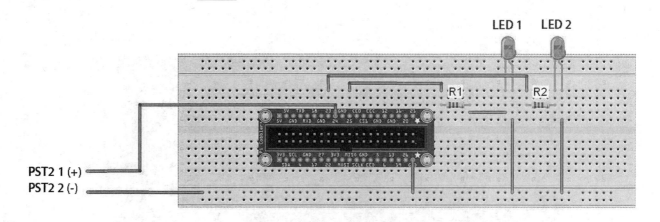

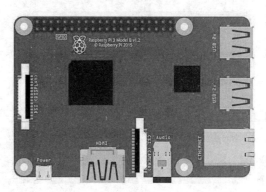

Figure 3-13 Fritzing diagram for test setup.

Table 3-1 Test Connections

Peripheral Connection	RasPi Physical Pin Number	RasPi BCM Pin Number
Resistor R1	22	25
Resistor R2	16	23
PST2 1(+)	18	24
PST2 2 (–)	14	GND
LED 1 cathode	14	GND
LED 2 cathode	14	GND

Modified Python Control Script

The Python control script that follows was modified from the *main.py* script that was presented previously. It has additional statements for the control of two additional GPIO pins, one to control a LED and the other to control the PST2. GPIO pin 23 controls the LED, whereas GPIO pin 24 controls the PST2. The key statement that allows the spoken data to be parsed is

```
speech_text = event.args["text"]
```

The Python string variable `speech_text` now contains all the user spoken text. This variable consequently can be tested to determine whether it contains specific instructions, which will then be acted on. For example, to turn on the AC lamp, the user must say, "OK Google, turn on the AC lamp." The following code snippet tests the string variable to determine whether it contains this phrase and, if so, activates the appropriate GPIO pin connected to the PST2:

```
if speech_text == 'turn on the AC lamp':
    GPIO.output(24, GPIO.HIGH)
```

Similar code statements detect phrases for controlling the LED and turning off the AC lamp.

The following listing is a completely modified *main.py* for the expanded control. Note that it is named *main_extended.py* on this book's companion website, www.mhprofessional.com/NorrisHomeAutomation, but must be renamed *main.py* and placed in the Home directory to operate correctly with the RasPi.

```python
#!/usr/bin/env python

from __future__ import print_function

import argparse
import os.path
import json

import google.oauth2.credentials
import RPi.GPIO as GPIO
from google.assistant.library import Assistant
from google.assistant.library.event import EventType
from google.assistant.library.file_helpers import existing_file

GPIO.setmode(GPIO.BCM)
GPIO.setup(23, GPIO.OUT)
GPIO.setup(24, GPIO.OUT)
GPIO.setup(25, GPIO.OUT)

def process_event(event):
    """Pretty prints events.
    Prints all events that occur with two spaces between each new
    conversation and a single space between turns of a conversation.
    Args:
        event(event.Event): The current event to process.
    """
    if event.type == EventType.ON_CONVERSATION_TURN_STARTED:
        print()
        GPIO.output(25,True)

    print(event)

    if (event.type == EventType.ON_CONVERSATION_TURN_FINISHED and
            event.args and not event.args['with_follow_on_turn']):
        print()
        GPIO.output(25,False)

    if event.type == EventType.ON_RECOGNIZING_SPEECH_FINISHED:
        speech_text = event.args["text"]
        print("speech text: " + speech_text)
        if speech_text == 'turn on the LED':
            GPIO.output(23, GPIO.HIGH)
```

```
        if speech_text == 'turn off the LED':
            GPIO.output(23, GPIO.LOW)
        if speech_text == 'turn on the AC lamp':
            GPIO.output(24, GPIO.HIGH)
        if speech_text == 'turn off the AC lamp':
            GPIO.output(24, GPIO.LOW)


def main():
    parser = argparse.ArgumentParser(
        formatter_class=argparse.RawTextHelpFormatter)
    parser.add_argument('--credentials', type=existing_file,
                        metavar='OAUTH2_CREDENTIALS_FILE',
                        default=os.path.join(
                            os.path.expanduser('/home/pi/.config'),
                            'google-oauthlib-tool',
                            'credentials.json'
                        ),
                        help='Path to store and read OAuth2 credentials')
    args = parser.parse_args()
    with open(args.credentials, 'r') as f:
        credentials = google.oauth2.credentials.Credentials(token=None,
                                                **json.load(f))


    with Assistant(credentials,"My Google Home device") as assistant:
        for event in assistant.start():
            process_event(event)



if __name__ == '__main__':
    main()
```

TIP: The above listing contains very long sets of instructions that display on multiple lines in the listing. The correct script requires that the long instructions sets be entered without line breaks and in accordance with Python's indention format. I highly recommend that you download the script from this book's website unless you are very comfortable in writing Python scripts.

Test Run

The modified *main.py* is run using the same shell script command previously described:

```
bash google-assistant-init.sh
```

I first tested the new LED with the spoken command, "OK Google, turn on the LED." I observed that the new LED did light up as expected. The original LED indicating an ongoing conversation also lit but stayed on for only the conversation duration.

I next spoke the phrase, "OK Google, turn off the LED." I then observed the new LED turn off, which confirmed that proper LED control was happening.

Similarly, speaking the phrases, "OK Google, turn on the AC lamp" and "OK Google, turn off the AC lamp," controlled the AC lamp through the PST2 peripheral, as it had in Chapter 2's demonstration. Of course, you can extend the PST2 range by using the RF devices I discussed in Chapter 2. You still need to modify the preceding script to accommodate the active low impulse required by the RF receiver using exactly the same code listed in the Chapter 2 Python script.

These simple demonstrations amply show that the RasPi Google Home device is now capable of controlling a variety of devices, including mains-powered devices.

Still More Extending of the RasPi Google Home Device

You probably have noticed by now that all the control actions implemented on the RasPi Google Home device have been binary. By this I mean that the device being controlled is either on or off. While this situation is perfectly acceptable for most HA applications, there are some applications where the device must be controlled in a more granular sense. I have developed an approach that allows for the direct interpretation of spoken commands into equivalent RasPi control actions beyond the simple on/off control states.

I will show how this may be accomplished in the next demonstration using a hobby-grade servo system. The servo I used is shown in Figure 3-14. It is a relatively inexpensive model yet is perfectly suited for this demonstration.

You might at this point question the use of a servo in an HA application. I would respond that servos are often used in home security applications to remotely position Web-based surveillance cameras. In any case, the primary purpose of this demonstration is to show you how to control an HA device beyond simply turning it on or off.

Figure 3-14 Hobby-grade servo.

Basic Servo Facts

It is very important to understand how a servo functions before attempting to create a program to control one. An analog servo is essentially an electric motor that incorporates an electronic control circuit that receives periodic digital pulses and positions or rotates the motor shaft in response to those pulses. Figure 3-15 is a block diagram illustrating the principal components that make up an analog servo.

The input digital pulse train that controls the servo's shaft position is known as a *pulse-width modulation* (PWM) waveform. The width of the high portion of the periodic pulse train is exactly proportional to the shaft position. Pulse widths range from 1.0 to 2.0 milliseconds (ms)

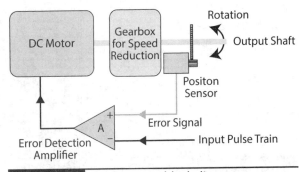

Figure 3-15 Analog servo block diagram.

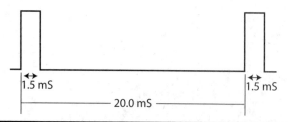

Figure 3-16   A 1.5-ms pulse-width, 50-Hz PWM waveform.

and correspond to ±60° shaft positions. A series of high-level pulses with widths of 1.5 ms and a frequency of 50 hertz (Hz) will cause the shaft position to be at its neutral position or midrange between the ±60° shaft end positions. Figure 3-16 shows a 50-Hz, 1.5-ms PWM waveform.

The Python software controlling the servo must generate a pulse width corresponding to the commanded position for the servo shaft position angle.

Test Setup

The Fritzing diagram in Figure 3-17 shows how the servo is connected to the existing demonstration hardware. You should notice that I used a separate 6-V power supply for the servo. Do not try to use the RasPi 5-V supply because there is insufficient current to drive the servo and the RasPi simultaneously. You will cause the RasPi to crash if you attempt to connect the servo power input to the 5-V supply. The 3.3-V GPIO pin output voltage is adequate to control the servo signal line without any issue.

Python Control Script Incorporating the PWM Option

The *Rpi.GPIO* library provides the means to easily generate PWM signals suitable for servo control. The essential servo control parameters

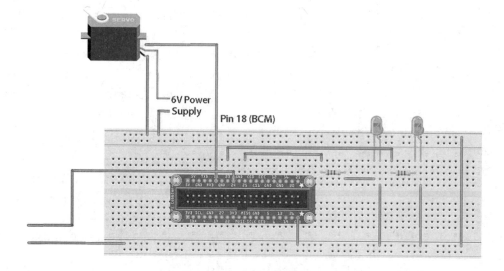
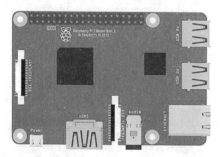

Figure 3-17   Fritzing diagram for servo test setup.

for the hobby servo used in this demonstration are listed in Table 3-2. Please note that the pulse width times for the hobby grade servo used in this demonstration are considerably longer than the standard pulse widths discussed above. This is not unusual and is primarily due to the inexpensive components used in such a device. You should always read the manufacturer's datasheet to ensure you are using the correct timing information for the device in use.

I adjusted the shaft position notation from the $\pm 60°$ ranges to a single $0–120°$ range to facilitate the spoken commands used with the RasPi Google Home device. The phrase "OK Google turn servo to zero" will actually cause the servo shaft to turn to the $–60°$ position. Likewise, the phrase "OK Google turn servo to one hundred twenty" will cause the servo shaft to turn to the $60°$ position. Finally, the phrase "OK Google turn servo to sixty" will position the servo shaft to the neutral or $0°$ position.

The following is the modified Python script incorporating the PWM control statements. It is named *main_servo.py* on this book's companion website but must be renamed *main.py* in the Home directory.

Table 3-2 Servo Control Parameters

Parameter	Value
Frequency	50 Hz (20-ms period)
Neutral position pulse width	7.5 ms
–60° position pulse width	5.0 ms
+60° position pulse width	10.0 ms
Power supply	6 V

```
#!/usr/bin/env python

from __future__ import print_function

import argparse
import os.path
import json

import google.oauth2.credentials
import RPi.GPIO as GPIO
from google.assistant.library import Assistant
from google.assistant.library.event import EventType
from google.assistant.library.file_helpers import existing_file

GPIO.setmode(GPIO.BCM)
GPIO.setup(18, GPIO.OUT)
GPIO.setup(23, GPIO.OUT)
GPIO.setup(24, GPIO.OUT)
GPIO.setup(25, GPIO.OUT)

# setup PWM output on pin 18 at a 50 Hz rate
pwm = GPIO.PWM(18, 50)
# start the PWM pulses for a 0 degree position
pwm.start(5.0)
```

```python
def process_event(event):
    """Pretty prints events.
    Prints all events that occur with two spaces between each new
    conversation and a single space between turns of a conversation.
    Args:
        event(event.Event): The current event to process.
    """
    if event.type == EventType.ON_CONVERSATION_TURN_STARTED:
        print()
        GPIO.output(25,True)

    print(event)

    if (event.type == EventType.ON_CONVERSATION_TURN_FINISHED and
            event.args and not event.args['with_follow_on_turn']):
        print()
        GPIO.output(25,False)

    if event.type == EventType.ON_RECOGNIZING_SPEECH_FINISHED:
        speech_text = event.args["text"]
        print("speech text: " + speech_text)
        if speech_text == 'turn on the LED':
            GPIO.output(23, GPIO.HIGH)
        if speech_text == 'turn off the LED':
            GPIO.output(23, GPIO.LOW)
        if speech_text == 'turn on the AC lamp':
            GPIO.output(24, GPIO.HIGH)
        if speech_text == 'turn off the AC lamp':
            GPIO.output(24, GPIO.LOW)
        if speech_text == 'turn Servo to 0':
            pwm.ChangeDutyCycle(5.0)
        if speech_text == 'turn Servo to 30':
            pwm.ChangeDutyCycle(6.25)
        if speech_text == 'turn Servo to 60':
            pwm.ChangeDutyCycle(7.5)
        if speech_text == 'turn Servo to 90':
            pwm.ChangeDutyCycle(8.75)
        if speech_text == 'turn Servo to 120':
            pwm.ChangeDutyCycle(10.0)
        if speech_text == 'turn Servo off':
            pwm.stop()

def main():
    parser = argparse.ArgumentParser(
        formatter_class=argparse.RawTextHelpFormatter)
    parser.add_argument('--credentials', type=existing_file,
                        metavar='OAUTH2_CREDENTIALS_FILE',
                        default=os.path.join(
                            os.path.expanduser('/home/pi/.config'),
```

```
                        'google-oauthlib-tool',
                        'credentials.json'
                    ),
                    help='Path to store and read OAuth2 credentials')
    args = parser.parse_args()
    with open(args.credentials, 'r') as f:
        credentials = google.oauth2.credentials.Credentials(token=None,
                                                    **json.load(f))

    with Assistant(credentials,"My Google Home device") as assistant:
        for event in assistant.start():
            process_event(event)


if __name__ == '__main__':
    main()
```

TIP: The above listing contains very long sets of instructions that display on multiple lines in the listing. The correct script requires that the long instructions sets be entered without line breaks and in accordance with Python's indention format. I highly recommend that you download the script from this book's website unless you are very comfortable in writing Python scripts.

Test Run

The modified *main.py* is run using the same shell script command described previously:

```
bash google-assistant-init.sh
```

I first tested the servo to move to the 60° position with the spoken command "OK Google, turn servo to sixty." I observed that the servo moved from the initial 0° position to the 60° position as commanded. The conversation LED also turned on, as expected, indicating that an ongoing conversation was in progress. Incidentally, the spoken response was "Sorry, I don't know how to help with that yet." This is the Google Assistant's standard phrase generated when dealing with an abnormal request such as turning on a servo. You should ignore the response. There is a way to suppress this warning response, but it would needlessly complicate the software without any real benefit.

I next spoke the phrase "OK Google, turn servo one hundred twenty." I then observed the servo turn fully clockwise to one of its limiting positions. The other limiting position in my context is the 0° command. I went through all the other positional commands, confirming that the servo did reposition as expected.

Finally, I spoke the phrase "OK Google, turn servo off." I observed that the servo stopped running, as can be seen by recognizing that it is no longer making any minute vibrations.

This last demonstration showed that a RasPi Google Home device is capable not only of binary control but also, when using appropriate programming extensions, of generating specific and unique device control actions. The use of this extended control approach is very efficient when compared with implementing a similar approach using Web-based requests, as was demonstrated in Chapter 2. In the latter approach, you have to create many additional applets that would issue granular servo control movements. Additionally, the RasPi Web server would require an equal number of conditional blocks to process the unique Web requests for each servo positioning request. This additional complexity is completely removed when using my approach as detailed in this chapter. The choice of which approach to use is totally up to you and relies on the specific HA application and how best to meet all system requirements.

Summary

This chapter's intent was to demonstrate how to design and build a RasPi emulator of a Google Home device. This design used only the RasPi with the addition of a USB microphone and USB speaker for speech detection and audio response.

I initially showed you how to set up the RasPi for proper audio operations that would support the design. A Python 3 installation including Google Assistant software was next introduced, including how to run a virtual Python environment.

The installation and configuration of a Google Cloud project was discussed next.

This mirrored in many respects the Chapter 1 presentation. The important authentication process was covered in fine detail.

I next presented a Python script that would run the Google Assistant software to cause the RasPi to emulate a Google Home device. A test was then conducted that proved that the RasPi did function properly as a Google Home device.

The following discussion included a demonstration of how to extend the default functions for the RasPi Google Home emulator. The test included controlling a LED and an AC lamp, similar to what was done in Chapter 2.

Finally, I presented an approach for device control that went well beyond the simple on/off controls demonstrated so far. This test showed how to control a servo motor, similar to one that might be used with a remote-controlled webcam.

## CHAPTER 4

Raspberry Pi GPIO Control with an Amazon Echo

IN THIS CHAPTER, I WILL demonstrate how to control some RasPi GPIO pins using an Amazon Echo personal voice assistant. This is similar to what was demonstrated in Chapter 2, except that in Chapter 2 I used the Google Home device. There is a significant difference between the approach used by Amazon and that by Google for connecting to voice assistant devices. I will explain the Amazon approach and point out some pros and cons of the Amazon and Google methods.

Wemo Demonstration

It would be informative and helpful to first demonstrate how the Amazon Echo connects with a commercial smart home device. This demonstration uses a Belkin-manufactured Wemo device that is directly controlled by an Echo unit. This demonstration uses a Wemo mini WiFi Smart miniplug, which is shown in Figure 4-1.

This device must first be set up using a smartphone app called the *Wemo app*. I used an iOS smartphone for app installation. Figure 4-2 shows the smartphone Wemo app screen.

The Echo device also must be linked to the Wemo device using this spoken command: "Alexa, discover devices." It typically takes an Echo device about 45 seconds to complete the discovery process. I had previously plugged an AC table lamp into the Wemo device in order to have an easy way to determine when the Wemo device was activated. I did see the table lamp turn on when I spoke this command: "Alexa, turn on the Wemo."

The Wemo device may also be controlled directly from the Wemo app simply by tapping

Parts List

Item	Model	Quantity	Source
RasPi 3	B or B+	1	adafruit.com amazon.com mcmelectronics.com
Smart plug	Belkin Wemo miniplug	1	amazon.com
Echo device	Any Alexa unit such as Basic, Spot, Show, or Tap	1	amazon.com
Power Switch Tail	II	1	amazon.com
AC table lamp	Commodity	1	Home improvement store

Figure 4-1 Wemo smart miniplug.

Figure 4-2 Wemo app screenshot.

on the On/Off button icon. Turning off the lamp with the Wemo was done using this command: "Alexa, turn off the Wemo." The lamp turned off after the command was spoken. These actions confirmed that the Echo unit was properly controlling the Wemo device through my local WiFi network. This was an important confirmation that directly leads into how the Echo device can control the GPIO pins on a RasPi. But first I need to introduce the Alexa Skill before describing how to control GPIO pins with an Echo device.

Alexa Skill

Alexa is the name of Amazon's voice service, just as *Google Assistant* is the name of Google's voice service. Alexa is also the alert name that activates voice service for an Echo unit, as I specified in the preceding spoken-command examples. Alexa provides certain functionalities, or *skills*, as they are preferably called in the voice service. Thousands of skills are presently available, with many provided by companies that manufacture devices for the Echo line of devices or provide complementary services that are integrated with overall Amazon services. Some examples of the latter include Starbucks, Uber, and Capitol One.

The Alexa Skills Kit (ASK) is a collection of APIs, utilities, documentation, and code samples that allow developers to fairly easily create and test new skills. Developers using ASK can take advantage of a powerful infrastructure in designing and building high-quality skills capable of being run on tens of millions of Alexa-enabled devices. ASK is Amazon's answer to Google's Voice Assistant API and Cloud Platform infrastructure. I will not comment on which is better or worse but will simply say that I am presenting both in this book for your information and use as you see fit.

There some basic components that make up an Alexa skill you should know about. These are shown in the custom Alexa Skill design screen in Figure 4-3.

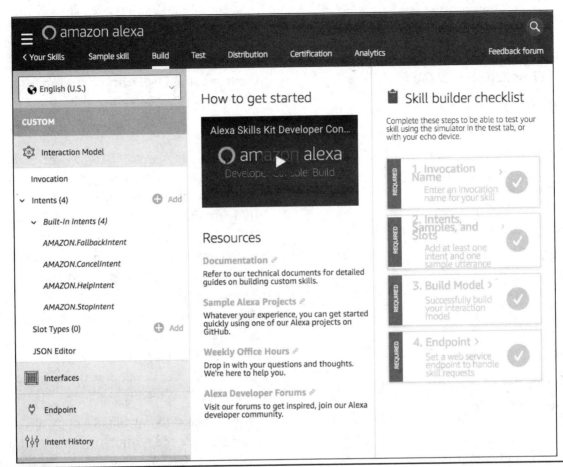

Figure 4-3 Custom Alexa Skill design screen.

The key skill components are

- **Utterance:** Any spoken phrase

- **Invocation:** A spoken phrase that starts an action (An action is also known as an *intent* in the ASK framework.)

- **Slot:** The objects of the intent, that is, what you want to do

- **Interaction model:** The collection of utterances, invocations, and slots that make up a conversation between a user and an Echo device

The spoken command "Alexa, turn on the Wemo" is an utterance that begins a conversation. The word *Alexa* is the default alert word to an Echo device, arming it to process the utterance that follows. The intent is specified by the phrase *turn on*, and the slot is *Wemo*, which is the intent's object. The intent and object are then transmitted to an Amazon server, which has previously been configured to recognize the Wemo object as well as various intents that could be applied to the object. If you try to apply an intent that was not previously set up, the Alexa service will respond, "Sorry, I am not sure" or "Sorry, I don't know that one" or some similar response phrase indicating that Alexa has no concept of the spoken intent.

Amazon has made available thousands of prebuilt skills for your immediate use. These skills come mainly from original equipment manufacturers (OEMs) and service providers that use Echo devices. The following is a list of the skills categories that I extracted from https://www.amazon.com/alexa-skills/b?ie=UTF8&node=13727921011 website:

- Business and finance
- Communications
- Connected cars
- Education and reference
- Food and drink
- Games, trivia, and accessories
- Health and fitness
- Home services
- Kids
- Lifestyle
- Local
- Movies and TV
- Music and audio
- News
- Novelty and humor
- Productivity
- Shopping
- Smart home
- Social
- Sports
- Travel and transportation
- Utilities
- Weather

However, no prebuilt skill is available for the Raspberry Pi in any of these categories. There is the Wemo skill provided by Belkin in the Smart Home category that I just demonstrated, and a clever developer named Nathan Henrie modified it so that the skill directly controls RasPi GPIO pins. He renamed the skill *Fauxmo*, which is a clever way of stating that the skill is a false, or faux, version of the original Wemo skill. The skill is directly downloadable from https://

github.com/n8henrie/fauxmo/, where *n8henrie* is Nathan's user name on Github. However, do not directly download this skill from the Github website because I will demonstrate an easier way to get it in the next section.

Fauxmo

I begin this section by stating that most of the instructions that follow came from an April 2017 Instructables blog, "Control Raspberry Pi GPIO Using Amazon Echo (Fauxmo)," written by Surendra Kane. I found the blog quite useful, although I did have to slightly modify some of the procedures to suit my situation. You will also find if you read the blog that Surendra actually modeled his blog after an earlier one dealing with the same subject. This is what is so great about open-source development: the tremendous support that is so readily available.

You first need to download a copy of the Fauxmo source code from Surendra's Github website using the following commands:

```
cd ~
sudo git clone https://github.com/
    kanesurendra/echo-pi.git
```

Next, move to the directory holding the Python script, and execute it with these commands:

```
cd echo-pi
python gpio_control.py
```

Now you can try asking your Echo device to discover any new devices by using this spoken command: "Alexa, discover devices."

The Echo device should provide a lengthy response starting with "Starting discovery. This will take 45 seconds...."

I found that this approach did not discover the RasPi. I next tried using the Alexa app on my smartphone and manually began a device discovery in the app's Smart Home devices

section. This approach was successful, with 12 new devices labeled "gpio15" through "gpio26" appearing in the app's screen, as shown in Figure 4-4.

I could now individually control the GPIO pins using the app or by speaking commands such as "Alexa, turn on gpio 22" and "Alexa, turn off gpio 22." The next step in the demonstration involves connecting external peripherals to some GPIO pins for a proof of performance, as was done in previous chapters.

Test Setup

The test peripherals consist of a LED and an AC lamp, which are controlled by a PST2. Figure 4-5 is a Fritzing diagram showing how they are connected to the GPIO pins.

GPIO pin 22 controls the LED, and GPIO pin 24 controls a PST2, which, in turn, controls an AC table lamp.

Figure 4-4   Alexa app with new GPIO devices.

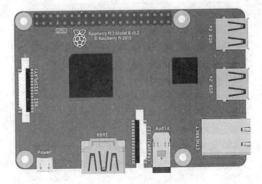

Figure 4-5   Test setup Fritzing diagram.

Test Run

These next spoken commands successfully controlled the LED:

"Alexa, turn on gpio 22."

"Alexa, turn off gpio 22."

Similarly, the following spoken commands successfully controlled the table lamp:

"Alexa, turn on gpio 24."

"Alexa, turn off gpio 24."

These commands are binary, meaning that the pin state is either on or off. This is because that is the only intent setup for any specific GPIO pin. Attempting to give some other action to a GPIO pin will only result in the usual "Sorry, I am not sure" or "Sorry, I don't know that one" Alexa responses. Figure 4-6 is a screenshot of the logger output instantiated by the control script.

You can see at the beginning of the log screen all the Fauxmo virtual devices being registered with attached port numbers. I will describe what these port numbers mean in a later section. The last few lines of log output show where I controlled first the LED attached to GPIO 22 and then the AC lamp (PST2), which is attached to GPIO 24. Note that the phrases *State True* and *State False* refer to the requested control action of "turn on" and "turn off," respectively.

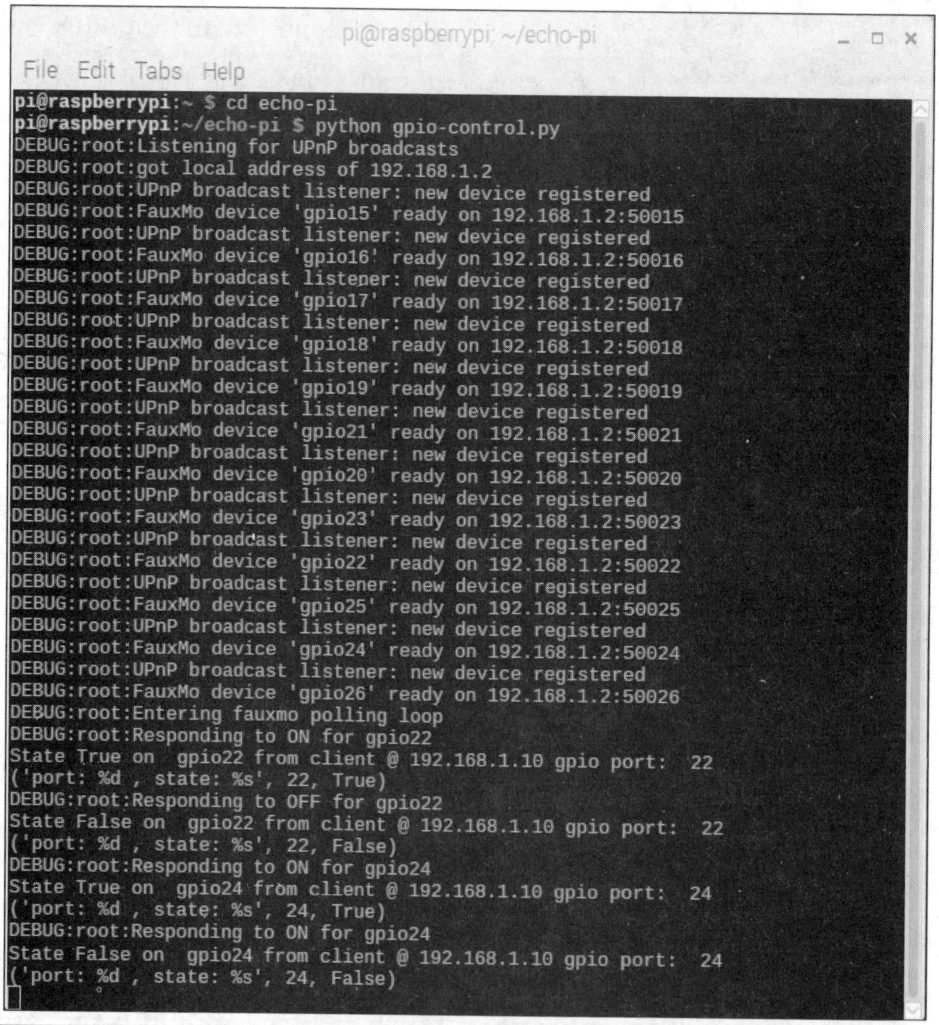

Figure 4-6 Log screen for Fauxmo server and control script.

Python Control Script

I think that it is informative to present the Python control script and discuss how it functions. Knowing more about it will likely allow you to further experiment and develop solutions more tailored to your specific needs.

```python
"""
  Author: Surendra Kane
  Script to control individual Raspberry Pi GPIO's.
  Applicable ONLY for Raspberry PI 3, based on schematics.
  Please modify for other board versions to control correct GPIO's.
"""

import fauxmo
import logging
import time
import RPi.GPIO as GPIO

from debounce_handler import debounce_handler

logging.basicConfig(level=logging.DEBUG)

GPIO.setmode(GPIO.BCM)
GPIO.setwarnings(False)

gpio_ports = {'gpio1':1,'gpio2':2,'gpio3':3,'gpio4':4,'gpio5':5,'gpio6':6,'gpio7':7,
  'gpio8':8,'gpio9':9,'gpio10':10,'gpio11':11,'gpio12':12,'gpio13':13,'gpio14':14,
  'gpio15':15,'gpio16':16,'gpio17':17,'gpio18':18,'gpio19':19,'gpio20':20,'gpio21':21,
  'gpio22':22,'gpio23':23,'gpio24':24,'gpio25':25,'gpio26':26}

class device_handler(debounce_handler):
    """Triggers on/off based on GPIO 'device' selected.
        Publishes the IP address of the Echo making the request.
    """
    """
    TRIGGERS = {"gpio1":50001,
                "gpio2":50002,
                "gpio3":50003,
                "gpio4":50004,
                "gpio5":50005,
                "gpio6":50006,
                "gpio7":50007,
                "gpio8":50008,
                "gpio9":50009,
                "gpio10":50010,
                "gpio11":50011,
                "gpio12":50012,
                "gpio13":50013,
                "gpio14":50014,
    """
    TRIGGERS = {"gpio15":50015,
```

```python
            "gpio16":50016,
            "gpio17":50017,
            "gpio18":50018,
            "gpio19":50019,
            "gpio20":50020,
            "gpio21":50021,
            "gpio22":50022,
            "gpio23":50023,
            "gpio24":50024,
            "gpio25":50025,
            "gpio26":50026}

    def trigger(self,port,state):
        print('port: %d , state: %s', port, state)
        if state == True:
            GPIO.setup(port, GPIO.OUT)
            GPIO.output(port,GPIO.HIGH)
        else:
            GPIO.setup(port, GPIO.OUT)
            GPIO.output(port,GPIO.LOW)

    def act(self, client_address, state, name):
        print "State", state, "on ", name, "from client @", client_address,
            "gpio port: ",gpio_ports[str(name)]
        self.trigger(gpio_ports[str(name)],state)
        return True


if __name__ == "__main__":
    # Startup the fauxmo server
    fauxmo.DEBUG = True
    p = fauxmo.poller()
    u = fauxmo.upnp_broadcast_responder()
    u.init_socket()
    p.add(u)

    # Register the device callback as a fauxmo handler
    d = device_handler()
    for trig, port in d.TRIGGERS.items():
        fauxmo.fauxmo(trig, u, p, None, port, d)

    # Loop and poll for incoming Echo requests
    logging.debug("Entering fauxmo polling loop")
    while True:
        try:
            # Allow time for a ctrl-c to stop the process
            p.poll(100)
            time.sleep(0.1)
        except Exception, e:
            logging.critical("Critical exception: " + str(e))
            break
```

The first thing you should notice is that this version of the script has activated only GPIO pins 15 through 26 as part of a TRIGGER array. I did read in the Fauxmo documentation that problems were encountered when too many GPIO pins were activated at a single time. This script allows you to activate GPIO pins 1 through 14 or 15 through 26 at any given time. You do that by uncommenting the desired TRIGGER array. Just ensure that the unwanted portion of the GPIO pins is also commented out.

Each GPIO pin has a unique port number that is assigned values contained in the active TRIGGER array. That port number, not the actual GPIO pin number, is what is sent to the Fauxmo server.

The Fauxmo server is the key software component that makes this whole project viable. A brief discussion regarding the Fauxmo server along with the underlying software communications protocol follows. You may skip reading the next section without any loss of continuity.

Fauxmo Server

I will like to first give credit to Chris, who publishes the Maker Musings blog. Apparently Chris is the open-source developer who created Fauxmo. The blog I am referencing is entitled, "Amazon Echo and Home Automation," and was published on July 13, 2015.

The Alexa service has had a Wemo skill for a relatively long time as compared with more recent HA skills. This skill was based on a WiFi-enabled AC mains power switch such as the one shown in Figure 4-1. Wemo devices use the UPnP communications protocol to "advertise" their availability in a local network. The UPnP Protocol is formally known as *Universal Plug and*

Play and was created to allow network devices to seamlessly interconnect with one another. These devices ordinarily include such things as PCs, printers, Internet gateways, WiFi access points, routers, and mobile devices. Additionally, the UPnP Protocol allows for the easy sharing of data among connected devices. UPnP uses the Internet Protocol (IP) as its backbone and also takes advantage of HTTP, SOAP, and XML technologies. Generally speaking, UPnP is best suited for consumer applications such as HA and is often avoided for use in business systems. This is because the protocol uses multicasting, which makes it "too chatty" for deployment in business or enterprise systems.

Figure 4-7 is a sequence diagram depicting how an Echo device communicates with a Wemo device using UPnP. I have found these diagrams to be very useful in explaining complex data communications configurations and interconnections.

Don't be scared if you find this figure a bit intimidating. It is basically just a timeline showing requests and responses between the Echo and Wemo units. The thing you should note that the same interaction is always happening between the units. I will shortly show you a figure with some actual data illustrating these requests/responses happening in real time between an Echo device and a RasPi. But first I want to continue with the Fauxmo story.

The Fauxmo software is an emulation of the original Wemo server software. The devices controlled in this emulation are known as *virtual Wemo devices* because they will respond as if they were physical Wemo devices but are something completely different, such as RasPi GPIO pins.

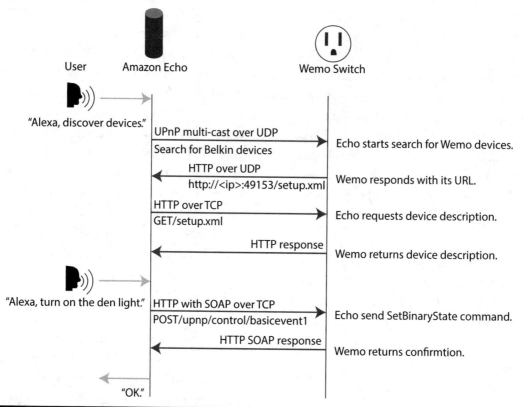

User   Amazon Echo   Wemo Switch

"Alexa, discover devices."

UPnP multi-cast over UDP   Echo starts search for Wemo devices.
Search for Belkin devices

HTTP over UDP
http://<ip>:49153/setup.xml   Wemo responds with its URL.

HTTP over TCP
GET/setup.xml   Echo requests device description.

HTTP response   Wemo returns device description.

"Alexa, turn on the den light."

HTTP with SOAP over TCP
POST/upnp/control/basicevent1   Echo send SetBinaryState command.

HTTP SOAP response   Wemo returns confirmtion.

"OK."

Figure 4-7   Echo/Wemo sequence diagram.

I am not going to discuss the complex functions of the Fauxmo server software other than to list them as they were provided in Chris's blog:

1. An IP address for each virtual switch

2. A listener for UDP broadcasts to address 239.255.255.250 on port 1900

3. A listener on port 49153 for each switch on its associated IP address

4. Logic to customize the search response and the setup.xml to conform to the UPnP Protocol and give the Echo the right information about each switch

5. Logic to respond to the on and off commands sent by the Echo and tie them to whatever action I really want to perform

Please read the blog if you want to delve deeper into the implementation details. The blog discussion is quite extensive and detailed, and many informative comments are also included with the discussion.

Surendra's control script imports the Fauxmo server package and sets up a polling service that constantly monitors the local network for any service requests emanating from a virtual Wemo device. The Python control script may be thought of as a wrapper to the Fauxmo software such that RasPi GPIO pins can be configured as virtual Wemo devices. Using this approach makes it quite easy to extend to other devices as desired.

Creating an Alexa Skill from Scratch

Up to this point in this chapter's discussions, I have relied on using an existing skill created by Belkin to communicate with Belkin-manufactured Wemo devices. This works and is quite useful to quickly implement a scheme to control GPIO pins on a RasPi. However, it would also be very useful to know how to design your own skill in case you want to communicate with a device that cannot emulate Wemo functionalities. This is why I will provide a detailed discussion on how to design and build an Alexa skill. This learning skill will involve a RasPi but not any GPIO pins. The skill itself is centered on the design of a simple memory game. The important point to note is that the skill-development process will be the same when designing for an HA application as it would be for implementing a simple game.

Memory Game Skill

The memory game skill I will be describing was created by John Wheeler, who also created the Flask-ASK package, which is a Flask extension that facilitates designing voice interfaces with the ASK. The memory game is a very simple application in which the Alexa asks you to repeat in reverse order three numbers that it spoke earlier. The application is implemented using Python script and runs on a RasPi. It uses the Flask package, which is a microframework written in Python that essentially supports Web-based applications. In fact, it is the most popular Python Web-development framework among open-source developers. I do not have the space in this book to delve into Flask but will state that many Internet tutorials are available, as well as several books completely devoted to Flask development.

You will first need to install the Flask-ASK package on the RasPi. As a gentle reminder, ensure that the RasPi OS has been updated and upgraded, as I have discussed previously in the book. Enter the following command to install Flask-ASK:

```
sudo pip install flask-ask
```

This command will take about a minute or so using a RasPi 3 (likely longer with a Model 2). After the package has been installed, open a nano editor and enter the following Python script. It is named *memory_game.py* and is also available for download from this book's website, www.mhprofessional.com/NorrisHomeAutomation.

```
import logging
from random import randint
from flask import Flask, render_template
from flask_ask import Ask, statement, question, session

app = Flask(__name__)
ask = Ask(app, "/")
logging.getLogger("flask_ask").setLevel(logging.DEBUG)
@ask.launch

def new_game():
    welcome_msg = render_template('welcome')
    return question(welcome_msg)
@ask.intent("YesIntent")

def next_round():
    numbers = [randint(0, 9) for _ in range(3)]
    round_msg = render_template('round', numbers=numbers)
    session.attributes['numbers'] = numbers[::-1]  # reverse
    return question(round_msg)
```

```
@ask.intent("AnswerIntent", convert={'first': int, 'second': int, 'third': int})

def answer(first, second, third):
    winning_numbers = session.attributes['numbers']
    if [first, second, third] == winning_numbers:
        msg = render_template('win')
    else:
        msg = render_template('lose')
    return statement(msg)

if __name__ == '__main__':
    app.run(debug=True
```

Flask-ASK also permits the separation of executable code from speech using templates. Therefore, you will need to start another nano editor session and enter the following speech responses into a file named *templates.yaml*. It, too, is available from this book's website.

```
welcome: Welcome to memory game. I'm going
to say three numbers for you to repeat
backwards. Ready?

round: Can you repeat the numbers
    {{ numbers|join(", ") }} backwards?

win: Good job!

lose: Sorry, that's the wrong answer.
```

The templates.yaml file must be located in the same directory as the *memory_game.py* script. I will assume that you will likely put them both in the Home directory.

If you run the script at this point, you will see that it starts a server on http://127.0.0.1:5000, which is the local host at port 5000. Unfortunately, the Alexa service has no way to directly access the local host, and the application cannot run successfully. Fortunately, there is a unique and clever answer to this problem, and its name is *ngrok*.

ngrok

ngrok is a utility that will open a secure tunnel to the local host, allowing the Alexa service to communicate directly with the RasPi. A secure tunnel is just an https URL that points to the internal Web server set up by Flask. There is no public host name, and you do not need to open any firewall protections. It is exactly what is needed for the skills deployment on a RasPi.

You will need to install ngrok, which is done by going to the website ngrok.com and following the simple set of instructions posted there. You can execute ngrok once it is installed by entering the following command:

```
./ngrok http 5000
```

You should see a screen similar to the one shown in Figure 4-8 after you execute the command.

The key bit of information that you eventually need is the https URL. The one shown in the figure is https://11941605.ngok.io and is also called an *endpoint*. Yours will be different. In fact, whatever URL is shown will not be used in the actual connection until you are ready to run the game. The URL created changes each time you start ngrok, and that particular URL must be the one entered into the skill prior to use. The next section discusses how to configure the Memory Game skill.

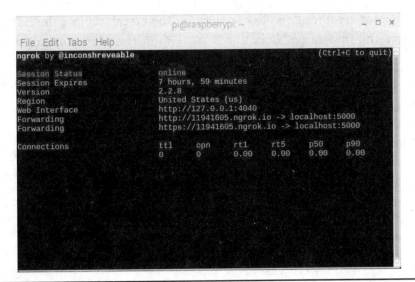

Figure 4-8   ngrok status screen.

Building and Configuring the Memory Game Skill

You will need to log onto your Amazon developer account to start the process of building an Alexa skill. Go to https://developer .amazon.com, and either login or create a new account by clicking on the appropriate button. It is fast, free, and absolutely required to proceed with this project. Once logged in, go to your list of Amazon skills. Of course, it will be empty if this is the first skill you are creating. The following steps must be closely followed in the order presented or you will likely not be successful in creating the skill.

1. Click on the Create Skill button. The screen shown in Figure 4-9 should appear.

2. Leave the Skill Type set on the Custom Interaction Model.

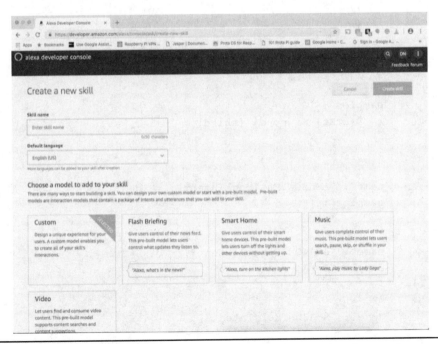

Figure 4-9   Initial screen for skill creation.

3. Enter "Memory Game" (without quotes) for both the Name and Invocation Name textboxes.

4. Copy the JSON file shown into the JSON editor window. There will already be four standard, prebuilt intents in the Editor window. The listed file includes these, so you can replace the entire existing Editor window contents with this file. This file is also available on this book's website, www.mhprofessional.com/ NorrisHomeAutomation, with the name *memory_game.json*.

```json
{
    "interactionModel": {
        "languageModel": {
            "invocationName": "memory game",
            "intents": [
                {
                    "name": "YesIntent",
                    "slots": [],
                    "samples": [
                        "yes"
                    ]
                },
                {
                    "name": "AnswerIntent",
                    "slots": [
                        {
                            "name": "first",
                            "type": "AMAZON.NUMBER"
                        },
                        {
                            "name": "second",
                            "type": "AMAZON.NUMBER"
                        },
                        {
                            "name": "third",
                            "type": "AMAZON.NUMBER"
                        }
                    ],
                    "samples": [
                        "{first} {second} {third}"
                    ]
                },
                {
                    "name": "AMAZON.FallbackIntent",
                    "samples": []
                },
                {
                    "name": "AMAZON.CancelIntent",
                    "samples": []
                },
                {
                    "name": "AMAZON.HelpIntent",
                    "samples": []
                },
```

```
        {
            "name": "AMAZON.StopIntent",
            "samples": []
        }
    ],
    "types": []
    }
}
```

5. Copy the following utterances into the Sample Utterances field:

```
YesIntent yes
YesIntent sure

AnswerIntent {first} {second} {third}
AnswerIntent {first} {second} and {third}
```

6. Next, run the ngrok utility and copy the https URL into the Endpoint field. Do not close the ngrok utility because the Alexa skill must use the URL Endpoint just generated.

7. Select the second radio button in the SSL Certificate settings. It states, "My development endpoint is a subdomain of a domain that has a wildcard certificate from a certificate authority."

8. Finally, you must build the interaction model by clicking on the Build button in the Alexa Skill console website. This will take a bit of time, so be patient.

The skill is ready for testing once it has been built and saved.

Test Run

Open another terminal window, and enter this command:

```
sudo python memory_game.py
```

This will start the Python script that was loaded earlier with a Web server running as local host at port 5000. However, the ngrok utility is still running and will have established a virtual tunnel between the Alexa service and the local host Web server. This is why it is so important to enter the current https URL created when you started the ngrok utility. Figure 4-10 shows a ngrok screen as it was running with a Memory Game session.

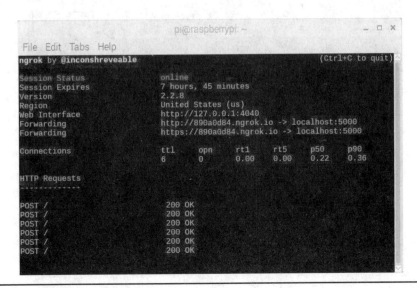

Figure 4-10 ngrok session in progress.

Now speak the following command to the Echo device: "Alexa, start the memory game." If all goes well, the response, should be "Welcome to the memory game. I'm going to say three numbers for you to repeat backwards. Ready?" Your response should be "Yes."

Alexa will then state three random numbers each between 0 and 9. You should repeat them back in reverse order. If you are successful, Alexa will respond, "Good job!" If unsuccessful, the response is "Sorry, that's the wrong answer." You may always change the Alexa responses by modifying the templates.yaml file (which contains the utterances).

There is an incredible amount of data flowing between the Alexa service and the RasPi, which you can see captured in the communications log shown in Figure 4-11.

Much of the incomprehensible text I believe is related to establishing and maintaining a secure tunnel between the Alexa service and the RasPi's local host Web server. In any case, if you do not see this data flow, then it likely means that a secure tunnel has not been created, and the Memory Game cannot work.

You deserve congratulations if you have successfully completed this demonstration. Creating an Alexa skill is moderately difficult, especially one that is implemented using a RasPi. The next demonstration takes the preceding demonstration one step further, where you will create a skill that will control any RasPi GPIO pin.

Figure 4-11 Communications log between the Alexa service and the RasPi.

Building and Configuring the RasPi GPIO Pin Control Skill

In this demonstration, you will be building a skill that will be able to turn on or off any RasPi GPIO pin. For convenience, I will be using the exact same physical setup shown in Figure 4-5, where GPIO pin 22 controlled a LED and GPIO pin 24 controlled an AC lamp using a PST2.

However, the skill is equally applicable to any available GPIO pin.

Python Control Script

Open a nano editor, and enter the following Python script. It is named *gpio_control.py* and is also available for download from this book's website.

```python
from flask import Flask
from flask_ask import Ask, statement, convert_errors
import RPi.GPIO as GPIO
import logging

GPIO.setmode(GPIO.BCM)

app = Flask(__name__)
ask = Ask(app, '/')

logging.getLogger("flask_ask").setLevel(logging.DEBUG)

@ask.intent('GPIOControlIntent', mapping={'status': 'status', 'pin': 'pin'})
def gpio_control(status, pin):

    try:
        pinNum = int(pin)
    except Exception as e:
        return statement('Pin number not valid.')

    GPIO.setup(pinNum, GPIO.OUT)

    if status in ['on', 'high']:    GPIO.output(pinNum, GPIO.HIGH)
    if status in ['off', 'low']:    GPIO.output(pinNum, GPIO.LOW)

    return statement('Turning pin {} {}'.format(pin, status))

if __name__ == '__main__':
    main()
```

Building the Alexa Skill

The same procedure detailed in the preceding Alexa skill-development discussion must be followed. I will not repeat all the steps except to detail the changes and modifications necessary for this particular skill build.

You need to name the skill after you click on the Create Skill button. I used the name *Raspberry Pi Control*, but you can name it anything you like.

The next item is to enter an invocation phrase. I used the phrase *pin control* because I felt that it would be natural and easy to say the phrase "Alexa, turn on pin control 22" to turn on GPIO pin 22.

Copy the JSON file shown into the JSON editor window. There will already be four standard prebuilt intents in the editor window. The listed file includes these, so you can replace the entire existing editor window contents with this file. This file is also available on this book's website with the name *raspicontrolintent.json*.

```json
{
    "interactionModel": {
        "languageModel": {
            "invocationName": "control pin",
            "intents": [
                {
                    "name": "GPIOControlIntent",
                    "slots": [
                        {
                            "name": "status",
                            "type": "GPIO_CONTROL"
                        },
                        {
                            "name": "pin",
                            "type": "AMAZON.NUMBER"
                        }
                    ],
                    "samples": [
                        "off",
                        "on"
                    ]
                },
                {
                    "name": "AMAZON.FallbackIntent",
                    "samples": []
                },
                {
                    "name": "AMAZON.CancelIntent",
                    "samples": []
                },
                {
                    "name": "AMAZON.HelpIntent",
                    "samples": []
                },
```

```
            {
                "name": "AMAZON.StopIntent",
                "samples": []
            }
        ],
        "types": [
            {
                "name": "GPIO_CONTROL",
                "values": [
                    {
                        "name": {
                            "value": "off"
                        }
                    },
                    {
                        "name": {
                            "value": "on"
                        }
                    }
                ]
            }
        ]
    }
}
```

At this point, you can save the skill and build it to check that it is viable, but you cannot deploy it until you set the Endpoint URL, as noted earlier.

Run ngrok, get the https URL, and enter it into the skill's Endpoint textbox. Also, ensure that you preset the proper SSL certificate, as discussed earlier. Now save and rebuild.

Test Run

The test run for this demonstration is run in exactly the same manner as the previous demonstration. Ensure that ngrok is still running, and then open another terminal window and run the Python control script using this command:

```
sudo python gpio_control.py
```

At this point, speak the following phrase: "Alexa, turn on pin control 22." You should now see the LED light up and hear the Alexa response: "Turning pin 22 on." Similarly, you can turn off the LED and turn the AC lamp on and off with the appropriate phrases.

This last demonstration should convince you that it is entirely possible to create an Alexa skill to control RasPi GPIO pins. I will say that having to set a URL prior to using the skill is a severe limitation that probably would exclude this approach for commercial HA applications. By contrast, using the Fauxmo approach, as was done in this chapter's second demonstration, is likely acceptable for a commercial HA application or your own efficient project.

NOTE: Chapter 5 was intended to be about how to build an Echo using only the RasPi in much the same way I wrote this chapter for the Google Home device. Unfortunately, after much development and trial and error, I was unable to build such a device. I do not know what to attribute the failure to except think that the Alexa voice service does not seem to be as accepting of open-source development as is Google's voice service. Hopefully, this will change in the future because voice services are very dynamic.

Summary

The chapter began with a demonstration of how to control a Wemo miniplug HA device using a smartphone app. I did this in preparation for introducing how the Amazon Alexa service functions and how to build a skill that could connect that service with a RasPi.

A short discussion regarding the Fauxmo server software followed. This software is an emulation of the Wemo software, which allows a RasPi to act as if it were a Wemo device. I provided a demonstration in which a LED and an AC lamp connected via a PST2 were controlled by the RasPi using spoken commands to an Echo device.

The ngrok utility was introduced, which makes possible the creation of a secure tunnel between the Alexa service and the RasPi Web server that is running as a local host. This utility was required for the follow-on demonstration.

A detailed discussion of how to create an Alexa skill from scratch followed. This skill involved building a simple Memory Game that was hosted on a RasPi. A user would start a conversation with an Echo device to run the game. This demonstration detailed all the necessary steps to build an Alexa skill.

Finally, another skill was built that directly controlled RasPi GPIO pins, which in truth controlled both the LED and AC lamp.

## CHAPTER 5

Home Automation Operating Systems

This chapter explores the subject of home automation operating systems (HA OSs). It probably seems strange to you to even know that there is such a thing as an HA OS. Most readers are aware of a computer OS, such as Windows 10 for PCs, OSX for Macs, and Linux for the open-source community. The most common and popular OS for the RasPi is Raspbian, which is a distribution within the Debian Linux family. It would be helpful to describe what a computer OS does before attempting to describe the purpose and function of an HA OS.

Computer Operating Systems

The Wikipedia definition of an OS system is as follows:

> An operating system (OS) is system software that manages computer hardware and software resources and provides common services for computer programs.

This definition is fairly accurate and is quite similar to those found in computer OS textbooks. I believe that the single most important item to know about an OS is that in itself it is just a program. Admittedly, it is a very large and complex program, but it is nonetheless just a program or, more specifically, a systems-level application. The OS loads automatically when the computer is powered on and eventually displays a friendly graphical user interface (GUI), which allows the user to easily control the whole system. It would be impossible for the average user to interact with a modern computer system without an OS functioning between the user and the hardware and software.

I am not going to delve into any OS specifics here because it is really not necessary for this section, and besides, it ordinarily requires taking

Parts List

Item	Model	Quantity	Source
RasPi 3	B	1	adafruit.com amazon.com mcmelectronics.com
Smart plug	Belkin Wemo miniplug	1	amazon.com
Smart lamp kit (includes bridge)	Philips Hue, white	1	amazon.com
Power Switch Tail	II	1	amazon.com

a one-semester college course to gain a good understanding of how a modern OS functions. The key point to understand about an OS is that it abstracts computer hardware and software to a point where relatively unsophisticated users can easily use a computer system to achieve their goals and objectives. These goals can be as simple as sending and receiving e-mail, using business applications, participating in social networks, and browsing the Internet. Likely 80 to 90 percent of all users match this description. The remaining portion of users is more knowledgeable and can develop applications as well as customize a computer system to meet their specific needs.

A modern OS has a lot of "moving parts" that are necessary for it to meet its goal of controlling a computer system based solely on mouse clicks on icons and small amounts of text input into browsers or dialog textboxes. This approach refers to the abstraction I mentioned earlier. Modern systems would be totally beyond most people's skills and patience without this abstraction layer present. It is not an understatement to say that the OS is responsible for today's computer revolution. Having an HA OS is similarly important for establishing and maintaining an HA revolution.

Home Automation Operating Systems

HA software is generally considered to be any type of software that is dedicated to controlling, configuring, or otherwise automating existing items and systems located in modern homes. These items and systems usually include large appliances such as stoves and refrigerators as well as small appliances such as coffee makers, media, and toasters. Home systems often include heating, ventilation, and air-conditioning (HVAC), irrigation, and lighting systems. The

tasks implemented in software normally include on/off and timed operations and notifying users when events happen. These events are usually triggered by sensors detecting an abnormal situation such as an unauthorized person in the home or a rapid ambient temperature change, which may be a precursor to an imminent fire.

HA software is often designed to work with multiple interfaces to the external environment, including relay closures, e-mail portals, and Extensible Messaging and Presence Protocol (XMPP), Z-Wave Protocol, and X10 Protocol products. The software structure is often a client-server architecture using a Web-based GUI or a mobile smartphone app. The ability to automate tasks is also built in using custom scripts and similar configuration files and utilities.

Huge concerns for HA device manufacturers and consumers are privacy and security. No one wants cyber intruders to enter private homes and business establishments. Determining who has access to vital systems that control home and business systems, which often record every moment and movement of users' lives, is of paramount importance. Users need strong assurances that automated devices are only communicating as designed and permitted and are never communicating or disclosing private information or data to unauthorized persons or organizations.

These security concerns are ample reason to use open-source solutions for HA applications. Understanding how an HA program functions is critical to ensuring that it is secure and will not expose you to all the threats that are ever-present whenever you are connected to the Internet. Unfortunately, some of the HA programs used with modern appliances and systems are proprietary and offer little to no understanding of how they are secured and how they can protect you from cyber threats.

The solution to this issue is to use an open-source protection layer in the form of a hub that ties all your devices together and provides a common user interface. This hub also should be designed for both security and ease of use and should be easy to customize to meet specific installations and requirements. This open-source hub is what I refer to as an HA OS, and many choices are currently available that will interoperate with a RasPi.

Open-Source HA OS Solutions

The following is a list of the most popular HA OS solutions currently available at the time of this writing. Of course, this is a very dynamic and fluid technology, and I am sure that some of these solutions will have disappeared by the time you read this. By contrast, new solutions will have also become available, so it will ultimately be up to you to research what is available and choose the solution that best fits your needs and objectives.

The following HA OS solutions are presented in alphabetical order, which means that you should impute no priority or preference based on their order in the list.

Calaos

This is a full-stack HA platform, which includes a server, touchscreen interface, Web interface, and mobile applications supporting both iOS and Android. It runs on a Linux platform. Calaos originated in France, so much of the support forums are in French, but the tutorials and supporting documentation have been translated into English.

This HA OS is licensed under GPL, version 3, and its source code is available from GitHub.

Domoticz

This HA system supports a wide range of devices, including uncommon devices such as weather stations and smoke detectors. Domoticz easily incorporates new communication protocols, which allows it to quickly integrate new devices into its hub. This capability is referred to as a third-party plug-in and is explained in great detail on the Domoticz website (www.domoticz.com). It uses HTML5 to implement its front-end, which makes it easily accessible using browsers and smartphone apps. It is of a lightweight design (meaning small memory footprint) and can be hosted on a RasPi.

It is written in C/C++ and is licensed under GPL, version 3. The source code is available from GitHub.

EventGhost

This is more like a utility program than a full-fledged HA OS. It is a home theater automation tool that is supported only on Microsoft Windows PCs. It allows users to control media PCs and attached hardware by using plug-ins that, in turn, trigger macros. Custom Python scripts also may be written to control the PC.

This HA OS is licensed under GPL, version 2, and its source code is available from GitHub.

Home Assistant

This HA OS is written in Python 3 and can run on any Linux platform that supports that language, including the RasPi. It also uses a Docker container for rapid and trouble-free deployment. This software easily integrates with IFTTT (If This, Then That) and Amazon services, providing for seamless interoperability with many devices. I will be introducing a practical demonstration of Home Assistant in this chapter.

Home Assistant is released to the public under the MIT license, and the source code is available from GitHub.

ioBroker

This is a JavaScript-based framework that can control lights, locks, thermostats, media, webcams, and much more. It will run on any platform that runs Node.js, which normally includes Windows, Linux, and macOS platforms.

ioBroker is released to the public under the MIT license, and the source code is available from GitHub.

Jeedom

This is a French-based open-source HA platform that controls lights, locks, media, and much more. It includes mobile apps for both Android and iOS. It operates on Linux PCs, which include the RasPi. The commercial company behind Jeedom also sells hardware hubs that provide ready-to-use solutions for setting up HA.

This HA OS is licensed under GPL, version 2, and its source code is available from GitHub.

LinuxMCE

This software is a media-based OS that runs on Linux and claims to interconnect all your home's media devices as well as HVAC, security, and telecommunications devices. It also claims to have the ability to run video games, which I suggest is problematic for a true HA OS–designed application.

It is released under the Pluto open-source license.

MisterHouse

This software uses Perl scripts, which enables it to run on all platforms, including Windows PCs, Macs, and Linux machines. It is voice and speech enabled and can provide many responses such as the current time and weather, warn of open doors and windows, announce phone calls, inform you if your child has been speeding, and so on.

openHab

The name openHab is short for Open Home Automation Bus and is probably the best known HA OS open-source software package used by HA developers and enthusiasts. This software is written in Java and will run on all popular OSs as well as on the RasPi. It supports hundreds of devices and is well suited to be adapted and modified to function with new devices because it is written in Java. openHab also supports device control using both iOS and Android apps. The openHab website (www.openhab.org/docs/concepts/) contains a very comprehensive introduction to the whole HA concept. I recommend that you take the time to read it. It will provide you with an excellent background in HA, well beyond what I can accomplish in this chapter.

This HA OS is licensed under the Eclipse public license, and its source code is available from GitHub.

OpenNetHome

This software will control lights, security devices, appliances, and so on. It's based on Java and Apache Maven and will operate on Windows, macOS, and Linux platforms including the RasPi.

This HA OS is licensed under GPL, version 3, and its source code is available from GitHub.

OpenMotics

This software is a bit different from other HA OSs described because it focuses mainly on individual device control and much less on the interoperability between devices. As such, it is hard to build a comprehensive HA OS using it, but it does have its place in providing a framework for device interfaces.

This HA OS is licensed under GPL, version 2, and its source code is available from GitHub.

Smarthomatic

This HA OS is an open-source framework that focuses on hardware devices and interface software rather than user interfaces. It is used for such things as controlling lights, appliances, and HVAC systems. It can measure ambient temperature and water house plants, provided that proper sensors and water systems are employed.

This HA OS is licensed under GPL, version 3, and its source code is available from GitHub.

Hass.io

Hass.io is a specialized HA OS that was created solely to ease the installation, configuration, and updating of the Home Assistant HA OS. You may properly consider it as a wrapper application for the Home Assistant because its purpose is to make the experience using Home Assistant as painless as possible while adding some new capabilities to the "wrapped" HA OS. These additional capabilities come in the form of plug-ins, which extend Home Assistant to use both Google Assistant and Let's Encrypt, as well as some other applications discussed later. Hass.io also allows the user to take Home Assistant configuration snapshots, which will allow for rapid and seamless restoration of an existing configuration.

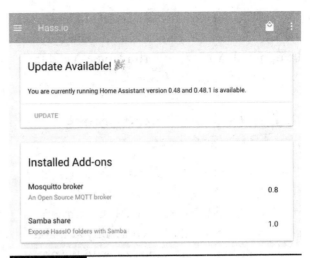

Figure 5-1 Hass.io dashboard.

Hass.io has been ported to run on the RasPi and is a great addition to support the Home Assistant HA OS, which can also be installed on the RasPi. A developer named Pascal Vizeli created Hass.io based on the ResinOS and Docker framework. This wrapper OS is configured using the Home Assistant user interface (UI). However, a Hass.io dashboard UI is also available, as shown in Figure 5-1, that will allow you to install add-ons to extend the Home Assistant capabilities.

The two add-ons shown in the figure are Mosquito broker and Samba share. The first one easily integrates with Google Assistant, whereas the second add-on makes the Home Assistant configuration accessible on a local network using the Samba/Windows application.

Other available add-ons that were not mentioned previously include

- Duck DNS
- Homebridge
- InfluxDB
- HASS Configurator
- AppDaemon

Hass.io is open source and released under the Apache 2.0 license. Hass.io is and will always be optional. You can still run Home Assistant wherever you can run Python.

Proprietary and Closed-Source Hardware/Software

For the sake of completeness, I have presented the following tables to show some of the proprietary and closed-source HA hardware/software solutions that are currently available. These will change, as is the case with the open-source solutions.

Proprietary Hardware

Table 5-1 details some of the more popular proprietary HA solutions. Listed in the table in most cases is the software that supports the manufacturers' devices.

Closed-Source Software

Table 5-2 details two closed-source HA solutions. The applicable computer platforms are also shown.

Table 5-1 Proprietary Hardware

Name	Configuration Utilities	Remarks
AMX LLC	Netlinx Studio, TP Design	Windows only
Control4	Composer	Uses a Linux kernel; configuration tools only work on Windows; platform also supports open hardware using the Z-Wave standard
Insteon	Insteon hub, Insteon for Windows	Lighting, appliances, sensors; mobile apps for Android and iOS; configuration tools only work with Windows
Lutron	—	Focused on lighting and shades; configuration tools only work on Windows
SmartThings	—	Lighting, appliances, sensors; mobile apps for Android and iOS
Vivint	—	Sensors and one-touch hardware for security

Table 5-2 Closed-Source HA Software Solutions

Name	Windows	macOS	Linux	Android	iOS	License	Remarks
Microsoft HomeOS	x					Academic	
HomeSeer	x	x	x	x	x		Bluetooth, 1-Wire, Z-Wave, X10, UPB, Insteon, Infrared

Installing the Home Assistant HA OS

I elected to demonstrate the Home Assistant HA OS because it is one of the most popular HA software packages, and its installation and configuration are very easy using the Hass.io software package introduced earlier. The only prerequisite for the installation is to use a RasPi 3, Model B, and not the recently introduced Model B+. I am not quite sure why it is mentioned in the installation other than there hadn't been sufficient time to try it in the installation process. That restriction may very well be removed by the time this book is published. In any case, just go to the website (https://home-assistant.io/getting-started) and carefully follow the eight steps described to install the Home Assistant software package. You will need a fresh 16- or 32-GB micro SD card as part of the installation. Ensure that it is a class 10 card to minimize the time it takes to write a new image onto the card.

There is a step in the installation process that will require you to edit an existing file on the micro SD card containing WiFi configuration details. This step is required to be accomplished before you insert the card into the RasPi. The edit is needed because the Hass.io automated installation procedure will automatically start to download the "real" Home Assistant software on the initial RasPi boot and will fail if there is no Internet connectivity. Of course, you can skip the edit step if you use a wired Ethernet connection.

The initial RasPi boot will take approximately 20 minutes, and an HA type icon will appear on the monitor screen for a few minutes. However, the icon will disappear, and there will be no indication of a successful installation. The monitor screen will be blank! Don't despair;

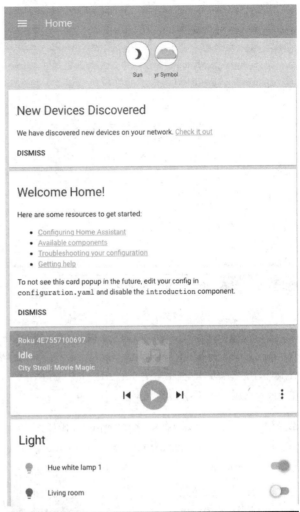

Figure 5-2 Initial Home Assistant welcome screen.

the installation has likely been successful. You must now use a browser on another computer connected to your local network and go to this site: http://hassio.local:8123/. You should see something similar to what is shown in Figure 5-2.

You likely noticed in the figure that the HA OS already discovered that I had a Philips Hue smart lamp connected as well as a Roku streaming device connected to my smart flat-panel TV. In fact, a show started appearing on the TV when I clicked on the Play button. Completing the smart lamp connection required only that I press a button on a Philips Hue bridge device. After that, I had full control of the smart light bulb, as you can see in Figure 5-3.

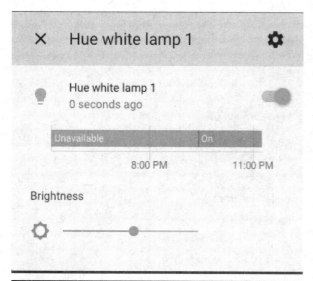

Figure 5-3 Smart light control.

The first thing you should do after connecting to the Home Assistant page is to add a password by editing the configuration file. Follow these steps to accomplish this task:

1. Open Home Assistant by going to http://hassio.local:8123/ using a browser on a computer connected to the home network.

2. Click on the Menu icon located in the top left-hand corner, and select Hass.io in the sidebar, which will appear.

3. Select the ADD-ON Store, which will appear in the Hass.io panel.

4. Install the HASS Configurator from the list of add-ons that appears in the store listing. You will be able to edit your Home Assistant configuration using a Web interface with the HASS Configurator.

5. Next, go to the Add-On Details page for the configurator, where you will be able to change settings, as well as start and stop the add-on. Follow the steps in the Details page to set up the add-on.

6. Set a password in the Config box, and don't forget to use quotes surrounding your password. The following listing is representative of what the initial configuration file will look like.

```
{
  "username": "admin",
  "password": "YOUR_PASSWORD_WITH_QUOTES",
  "certfile": "fullchain.pem",
  "keyfile": "privkey.pem",
  "ssl": false,
  "allowed_networks": [
    "192.168.0.0/16",
    "172.30.0.0/16"
  ],
  "banned_ips": [
    "8.8.8.8"
  ],
  "banlimit": 0,
  "ignore_pattern": [
    "__pycache__"
  ],
  "dirsfirst": false,
  "enforce_basepath": false
}
```

7. Click on SAVE to save your new password.

8. Next, click on START at the top of the Configurator add-on panel.

9. You will now be able to click the OPEN WEB UI link to open the Web UI in a new window.

Sign in

http://hassio.local:3218

Your connection to this site is not private

Username

Password

Cancel   Sign In

Figure 5-4   Login dialog box.

10. Type your username and password that you recently created in the dialog box, as shown in Figure 5-4. A Web-based editor should now appear, as shown in Figure 5-5. You can modify the Home Assistant configuration using this editor. Do not close this Web page because in the next section I discuss how to make some modifications to the Home Assistant HA OS using the Web editor.

Modifying the Home Assistant Using the Configurator

The first configuration task I will demonstrate is how to add a new icon into the Home Assistant sidebar, which when clicked will start the Configurator. The following code must be entered into the *configuration.yaml* file in order to add the new icon:

```
panel_frame:
  configurator:
    title: 'Configurator'
    icon: mdi:wrench
    url: 'http://hassio.local:8123'
```

Click on the Browse Folder icon in the upper left-hand corner of the Web editor to open a list of Home Assistant files that can be edited. Select the *configuration.yaml* file, as shown in Figure 5-6.

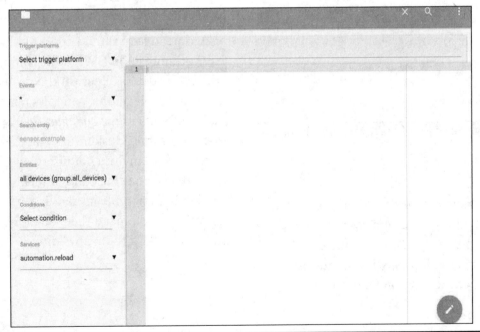

Figure 5-5   Configurator Web-based editor window.

select this file

Figure 5-6   Selection of the *configuration.yaml* file.

effect. I found that I had to power cycle the RasPi to reboot the system. Trying to reboot the RasPi over the Web was ineffective and a wasted effort. You should wait several minutes before attempting to log back into the Home Assistant.

Figure 5-7 shows the new welcome screen after a successful reboot. You can clearly see the new Configurator "wrench" icon in the sidebar, which was not present prior to the preceding configuration file edit.

Editing the Configuration Using Samba

It is also possible to avoid using the Configurator Web-based editor by installing the Samba add-on. Samba is a Windows application that allows Linux/Unix programs to interoperate with a Windows OS over a network. It is installed in the Home Assistant in the same manner that the Configurator was done. Go into the ADD-ON store and select the Samba share application for installation. Click on the Start button once the Samba application has been installed.

A Hass.io icon should appear in the Networking tab on a Windows computer connected to a local network. Now use any text editor available on the Windows machine to make the edits to the *configuration.yaml* file, as discussed earlier. Just ensure that the Windows editing program does not insert any special or unique formatting codes, which will render the configuration file invalid. In this context, I strongly recommend that you do not use Microsoft Word to do any editing.

Enter the listing at the end of the file, playing particular attention to the indentations in the listing. Save the newly edited file, and exit the Configurator. The Home Assistant must now be rebooted for the new configuration to take

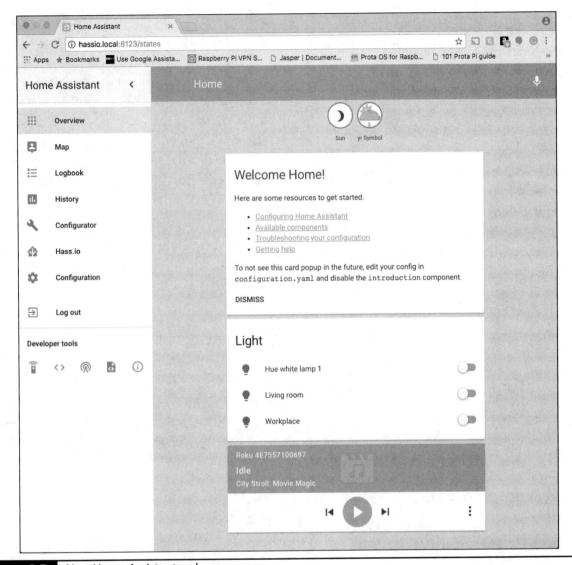

Figure 5-7 New Home Assistant welcome screen.

Configuring Integrations

It is now time to show you how to manually configure devices and services. Most HA devices and services will have dedicated instruction regarding how to integrate them into an HA OS. The first step, in most cases, is to locate the device/services manufacturer in the Home Assistant component Web page. This will be at www.home-assistant.io/components/. I chose adding the Belkin miniplug as a simple example for a device integration. Going to the URL shown earlier will reveal the Web page shown in Figure 5-8.

You may be a bit startled to find out that more than 1,100 devices and services are referenced by this Web page, which you can see by the total count (All) in the sidebar. I next clicked on the Belkin Wemo icon, and it took me to the Web page that describes how to integrate a Belkin Wemo unit into the Home Assistant. The following discussion is based strongly on the contents from that page, which I like because it explains in simple but clear terms how to integrate a Wemo device.

The Wemo component is used to integrate various Belkin devices with Home Assistant.

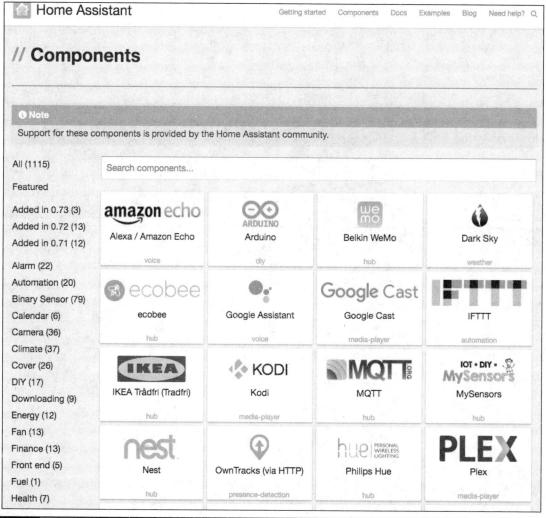

Figure 5-8   Initial Home Assistant components Web page.

Normally, Wemo devices will be automatically discovered if the Home Assistant discovery component service is enabled. However, manually loading the Wemo component will force a scan of the local network for Wemo devices, even if the discovery component service is not enabled.

Every discovered device results in an additional entry into the *configuration.yaml* file, for example, *configuration.yaml* entry

```
wemo:
```

Alternately, Wemo devices that do not seem to be discoverable may be statically configured. This may be the case if you have Wemo devices on a subnet other than the one where Home Assistant is running or have devices set up in a remote location that is reachable over a virtual private network (VPN). This approach is also useful if you wish to disable discovery for some Wemo dervices, even if they are local. An example of a static configuration is

```
wemo:
  static:
    - 192.168.1.23
    - 192.168.52.17
```

Note that any Wemo devices that are not statically configured but still reachable via the discovery service will still be added automatically to the Home Assistant configuration file. Also note that if you employ static IP addresses, you may need to set up your router (typically running the Dynamic Host Configuration Protocol [DHCP] service) to force the Wemo devices to use a static IP address. You should check the DHCP section of your router configuration for this capability.

If the device doesn't seem to work and all you see is the state "Unavailable" on your dashboard, check that your firewall doesn't block incoming

requests on port 8989 because this is the address to which the Wemo devices send their updates.

Emulated Wemo Devices

There are a number of software packages that emulate Wemo devices and often use alternative ports. All static configurations must include the port value, as shown in the next example.

```
wemo:
  static:
    - 192.168.1.23:52001
    - 192.168.52.172:52002
```

Remember that I discussed the Fauxmo Wemo emulation in Chapter 4. In that demonstration, I used port numbers ranging from 50015 to 50026 along with the RasPi IP address to control individual GPIO pins.

Automating the Home Assistant

Creating automating scripts is a key feature in successfully and effectively using an HA OS. The process of creating a script is called *automation*, and there a few basics that I need to cover for you to have a good understanding of this process.

Let's say that Mary arrives home in the evening and desires to have the hallway and living room lights turned on by the Home Assistant. This scenario may be broken down or analyzed as follows:

(trigger)	Mary arrives home
(condition)	in the evening
(action)	Turn on hallway and living room lights

This scenario has three distinct parts: the trigger, the condition, and the action. Each part is part of an automation rule and has a distinct role. These roles are

- **Trigger.** Describe event(s) that should trigger the automation rule. In this case, "Mary arrives home." The Home Assistant would need to be aware of the state change of Mary from "not being home" to "being home."

- **Condition.** A condition is an optional test that can limit an automation rule to function only for a specific requirement in your use case. The condition detects current system states. These states might include the current time, weather, sensors, people, and other things such as the sun. In this particular case, the rule should be acted on only after the sun has set. Time would not be a good test because sunset changes constantly. Multiple conditions also may need to be met before the rule is acted on.

- **Action.** The action will be performed after a rule is triggered and all conditions are met. In this example, both the hallway and living room lights are turned on after Mary arrives home in the evening. Actions can be a multitude of things including setting the temperature on a smart thermostat or activating a scene, which I describe below.

It is important to differentiate between the various roles. A light being on could very well be a condition or even a trigger, whereas turning a light on likely would be an action. Proper role identification is one key to creating workable automation rules.

Internal HA OS States

Automation rules depend on knowing the internal state of the HA OS. In the Home Assistant, the current state is available by clicking on a Developer Tool icon in the sidebar shown in Figure 5-9.

Figure 5-9   Icon for current state(s).

All the current HA OS states will be shown as entities. These entities can be anything, ranging from lights to people, even the sun itself. A state has multiple parts that are described in Table 5-3.

Table 5-3   State Components

Name	Description	Example
Entity ID	Unique identifier for the entity	light.kitchen
State	The current state of the device	home
Attributes	Extra data related to the device and/or current state	brightness

Changes in state may be used as trigger sources, and the current state can also be used in setting conditions.

HA OS Services

All Home Assistant actions are carried out using services. In the Home Assistant, all current services are available by clicking on a Developer Tool icon in the sidebar shown in Figure 5-10.

Figure 5-10   Icon for current services.

Services allow the HA to control anything already integrated into the OS, for example, turning on a light, running a script, or enabling a scene. Every service has both a domain and a name. For example, the service `light.turn_on` is capable of turning on any light registered in the HA system. Services can also pass parameters or arguments, which can set a specific color for a multicolored lamp and/or set a light intensity.

You will have to set an initial state in your automations in order for Home Assistant to enable them on restart. The following is an example of a code snippet to be placed in the *configuration.yaml* file to turn on automation:

```
automation:
- alias: Automation Name
  initial_state: True
  trigger:
...
```

Automation Example

It would beneficial at this point to actually create a working automation example. The automation rule is

Turn on a light after sunset

This rule implicitly defines a trigger that somehow tracks the sunset and will fire the rule when the sun does set. The service is `light.turn_on`, and if it is called without any parameters or arguments, it will turn on all integrated lights. An example automation rule entry in the *configuration.yaml* file would be

```
# Example configuration.yaml entry
automation:
   alias: Turn on the lights when the sun
sets
   initial_state: True
   hide_entity: False
```

Automation

≡ Check the battery state of a tablet

≡ Send notification if switch is used

≡ Turn on light when switch is used

≡ Update Available Notifications

Figure 5-11 Sample automation rules control panel.

```
trigger:
  platform: sun
  event: sunset
action:
  service: light.turn_on
```

Starting with Home Assistant, version 0.28, all the automation rules can be controlled with the front-end. Figure 5-11 shows a sample automation rules control panel. Using a control panel in the front-end means that automation rules can be reloaded (or unloaded) without restarting Home Assistant itself.

If you don't want to see the automation rule in your front-end, use the statement `hide_entity: True` to hide it. You can also use the statement `initial_state: 'off'` or `'false'` so that the automation is not automatically turned on after a Home Assistant reboot.

I added the preceding automation rule to the *configuration.yaml* file and rebooted the RasPi in order to reboot the Home Assistant OS with the modified configuration file. After the Home Assistant was started, I observed that the Hue lamp turned on after I ensured that the sun entity was false or off. This meant that it was sunset.

Modifications to the Automation Example

Let's suppose that after a few days of using the automation rule you observe that the lights went on after it was already dark and also that the lights went on when nobody was home. This means that it is time for some modifications and conditions to be added to the automation rule. One modification would be to add a time offset to the sunset trigger as well as adding a condition to test whether anyone is home. The following modified rule should replace the existing automation rule in the *configuration.yaml* file:

```
# Example configuration.yaml entry
automation:
    alias: Turn on the lights when the sun
        sets
    trigger:
      platform: sun
      event: sunset
      offset: "-01:00:00"
    condition:
      condition: state
      entity_id: group.all_devices
      state: 'home'
    action:
      service: light.turn_on
```

I did not test this particular automation script because I had not yet set up presence detection, which I describe in a later section. The preceding script is not the end of this tale of automation. You recently discovered that the bedroom lights also went on when the living room and kitchen lights went on after sunset. What you really want is for only the living room lights to turn on after sunset.

The first thing to do is to check the names of all the light entities that are integrated into the HA system. You can do this by clicking on the Developer Current State icon, Figure 5-9

in the Home Assistant sidebar. Write down the names of all the light entities. Suppose that they are `light.table_lamp`, `light.bedroom`, and `light.ceiling`. A group will be set up in the automation rule that will avoid the issue of hard coding the light entity IDs in the automation rule. This approach will separate the living room lights from all other lights in the automation rule.

The *configuration.yaml* file is once again modified with this new group to ensure that only the desired light(s) in the group are activated:

```
# Example configuration.yaml entry
group:
    living_room:
      - light.table_lamp
      - light.ceiling

automation:
    alias: Turn on the light when the sun
        sets
    trigger:
      platform: sun
      event: sunset
      offset: "-01:00:00"
    condition:
      condition: state
      entity_id: group.all_devices
      state: 'home'
    action:
      service: light.turn_on
      entity_id: group.living_room
```

The holidays are approaching, and you decide to purchase a remote control switch to control the Christmas tree lights using the Home Assistant. You first integrate the remote switch with the Home Assistant and find the appropriate entity ID using the State Developer tool. In this case, it is `switch.christmas_lights`. The automation rule must be modified once more to handle the switch, but the action `light.turn_on` is no longer applicable to the new device, which is a switch. Fortunately, Home Assistant has a service named `homeassistant.turn_on` that is capable of turning on any entity. The modified automation rule is

```
# Example configuration.yaml entry
group:
   living_room:
      - light.table_lamp
      - light.ceiling
      - switch.christmas_lights

automation:
   alias: Turn on the lights when the sun
      sets
   hide_entity: True
   trigger:
      platform: sun
      event: sunset
      offset: "-01:00:00"
   condition:
      condition: state
      entity_id: group.all_devices
      state: 'home'
   action:
      service: homeassistant.turn_on
      entity_id: group.living_room
```

Setting Up Presence Detection

Presence detection detects whether people are at home, which is a very valuable input for creating automation scripts. Knowing who is at home or where they are will open a whole range of automation options, including

- Sending a notification when a child arrives at school

- Turning on the air-conditioning when I leave work

The Home Assistant device tracker component provides presence detection. It supports three different methods for presence detection:

1. Scan for connected devices on the local network.

2. Scan for Bluetooth devices within range.

3. Connect to a third-party service.

Scanning for connected devices is easy to set up with options that include using Home Assistant components known as router-based devices/services, a portion of which is shown in Figure 5-12.

The second approach uses the Nmap utility. The Web page www.home-assistant.io/components/device_tracker.nmap_tracker/ describes how to set up the Nmap utility on the RasPi. This approach does have its limitations, but it will only be able to detect whether a device is at home, and modern smartphones may show as not home inaccurately because they disconnect from WiFi if idle after a certain time period.

You can also scan for Bluetooth and Bluetooth LE (light energy) devices. Fortunately, modern smartphones don't turn off Bluetooth automatically, although the range is smaller than with WiFi.

Home Assistant currently supports many third-party services for presence detection, such as OwnTracks over MQTT, OwnTracks over HTTP, GPSLogger, and Locative. Overall, a wide range of options is available, both for scanning the local network and for third-party services.

Home Assistant will know the location of your device if you are using a device tracker that reports a GPS location (such as OwnTracks, GPS Logger, the iOS app, and others). You will also be able to add names to the locations of devices by creating zones. In this way, you can easily locate on the state page where the people

Figure 5-12 A few of the Home Assistant presence-detection devices/services.

in your house are and use it both for triggers and conditions for automation scripts.

There is one caution in that if you're looking at the map view, then any devices in your home zone won't be visible. The Home Assistant map view is part of the front-end and is designed to display the location of all tracked devices, except those located in the home zone. You add the map by including the following simple entry in the *configuration.yaml* file:

```
# Example configuration.yaml entry
map:
```

Summary

This chapter described what an HA OS is and how it operates. I began it with a brief description of how a regular computer OS works and proceeded to extend that discussion to how the HA OS functions and how it differs from a computer OS.

A detailed list regarding a dozen open-source HA OSs was next presented. I also included a discussion of the Hass.io OS, which is a wrapper-type OS for the Home Assistant HA

OS addressed throughout the remainder of the chapter.

A through discussion on how to install and configure the Home Assistant HA OS on a RasPi 3, Model B, was next. A brief demonstration of using the Home Assistant with a Philips Hue smart bulb also was included.

I next showed how to modify the Home Assistant configuration file using a Web-based editor provided by a Home Assistant add-on named the Configurator. A Belkin Wemo miniplug was manually integrated into the Home Assistant using the Configurator.

I next demonstrated how to create an automation script that used both states and services. An extensive discussion followed about modifying the automation script three times to optimize its operation. The chapter concluded with a brief discussion of presence detection and how to implement it in the Home Assistant.

## CHAPTER 6

Z-Wave and Home Automation

USING Z-WAVE IN A RASPI HA project is the intriguing subject for this chapter. I have mentioned Z-Wave in previous chapters because it is a predominant and proven technology that has been adopted by many HA device manufacturers. These companies include established brands such as GE, Black & Decker, Schlage, ADT, and Draper.

Z-Wave Fundamentals

Z-Wave is a wireless packet-based RF communications platform. It is somewhat similar to WiFi because the latter also uses RF and packets, but there are also significant differences. Z-Wave uses RF transceivers that operate at 908.42 megahertz (MHz) in the United States and at 860 MHz in Europe. These frequencies belong to the industrial, scientific, and medical (ISM) radio band, which is significantly separated from WiFi frequencies, which are either at 2.4 or 5.0 gigahertz (GHz). The ISM band is often subject to much less noise and interference than the higher frequency WiFi bands. A Z-Wave RF transceiver has a typical range of 100 meters (m), which is comparable with a WiFi node. It is also subject to the signal attenuation that occurs indoors owing to walls

Parts List

Item	Model	Quantity	Source
RasPi 3	B	1	adafruit.com amazon.com mcmelectronics.com
Z-Wave duplex outlet	GE Z-Wave Plus Smart Lighting and Appliance Control Receptacle Outlet	1	amazon.com
Z-Wave controller	GE Z-Wave Wireless Lighting Control LCD Remote	1	amazon.com
Z-Wave outdoor module	Jasco Z-Wave Plug-In Outdoor Module (45704)	1	amazon.com
Z-Wave USB dongle	Aeon Labs DSA02203-ZWUS Z-Wave Z-Stick Series 2 USB Dongle	1	amazon.com

and doors interfering with the signal strength. Z-Wave signal extenders are available, just as there are WiFi signal extenders.

A group of Z-Wave transceivers makes up a *mesh*, with each transceiver considered a *node* in that mesh. Each node can both receive and transmit digital packets. Nodes not only can originate data packets but also can receive and retransmit data packets from other nodes. This action is known as *digipeating* and is a very important property of a mesh network. A node that is digipeating is called a *hop*. Only four hops are allowed for any specific data packet because of regulatory concerns regarding unlicensed operation within the ISM band. All data packets are destroyed after the fourth hop in a process known as *hop kill*.

Z-Wave Network Basics

The Z-Wave data communications protocol is in compliance with the Open System Interconnection (OSI) seven-layer network model. The complete seven-layer model is shown in Figure 6-1.

However, not all the model layers are implemented in the Z-Wave communications protocol. This is because the OSI model's seven

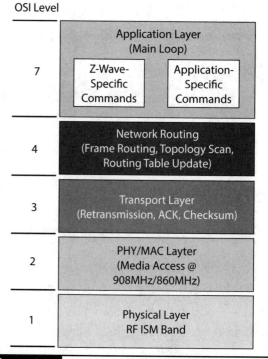

Figure 6-2 Z-Wave-specific network model.

layers were designed to cover a host of different protocols, some of which are irrelevant for Z-Wave. Figure 6-2 shows the four layers that are implemented with remarks in each layer detailing the implementations.

Layer 1 is the physical layer, which uses an ISM band RF transceiver to both send and receive digital packets. These packets are raw in nature in the sense that they are either decoded in the model upper layers for receive packets or simply sent out for the already formatted transmit packets from the upper layers. Every node has Carrier-Sense Multiple Access with Collision Detection (CDMA/CD) hardware, which determines when it is safe to transmit over the network. CDMA/CD has been compared with the very old-fashioned telephone party line, where a user would first pick up the headset and listen to determine if anyone was currently talking on the line. The user would then start the call if nothing was heard and otherwise hang up and try a little bit later. CDMA/CD

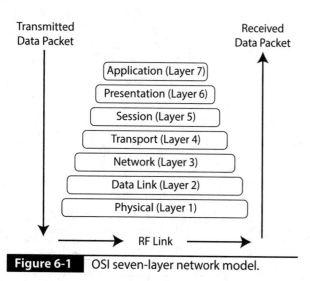

Figure 6-1 OSI seven-layer network model.

does the same except that the time frame is in milliseconds rather than minutes, as would be the case for a party line.

Layer 2 is the data link, where a packet to be transmitted has a synchronization preamble along with a start-of-frame (SoF) byte prepended to the data payload sent from higher OSI levels. Additionally, an end-of-frame (EoF) byte is appended to the end of the data payload. Received packets are treated in the reverse manner; that is, the received raw data packet has its framing bytes removed before the data payload is sent upward to the higher OSI layers.

Layer 3 is the transport, where additional bytes are added to the data packet being prepared for transmit. These bytes depend on the active communication process. Z-Wave is a connection-type network, which has a very robust method of ensuring packet delivery. Layer 3 uses an acknowledgment (ACK) byte and negative acknowledgment (NACK) byte to maintain a continuous communications link. A node acting as a receiver will send back an ACK packet to the transmit node if it successfully receives a data packet. Likewise, it will send back a NACK if the packet was corrupted or somehow not completely received. This *handshake* will continue until the original packet is successfully sent or a preset limit of retries is reached, a so-called timeout operation. Using ACKs and NACKs depends on detecting

errors in the data packet, which is why there are two checksum bytes included in every data packet. These bytes are used by the layer 3 implementation code to test the received data payload using a mathematical algorithm called the *cyclic redundancy check* (CRC). It should be apparent that layer 3 works mainly with receive packets to ensure a stable communications link. It must also generate the checksums for the transmit packets.

Layer 4 is for routing, where transmit packets have address information included in the packet to ensure that the packet arrives at the intended node. This layer also ensures that nodes will digipeat. Routing information is maintained in a table located in the primary network controller node. It is possible to have multiple primary controllers present in the network, but only one can be active at any given time. Figure 6-3 shows a simple network topology along with a routing table for that topology.

You should be able to see that there is no direct connection between nodes 4 and 5. Any packets exchanged between the two must go through node 2 or the longer path from node 2 to node 3 to node 6. Note there is an even longer path—node 1 to node 2 to node 3 to node 6—but this path would never be taken because it involves four hops, and the packet would be dumped because of the four-hop restriction I mentioned earlier.

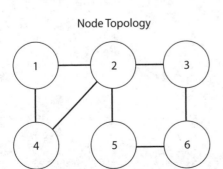

Node Topology

Routing Table

	1	2	3	4	5	6
1	0	1	0	1	0	0
2	1	0	1	1	1	0
3	0	1	0	0	0	1
4	1	1	0	0	0	0
5	0	1	0	0	0	1
6	0	0	1	0	1	0

Figure 6-3 Simple mesh network topology with routing table.

Layer 7 is the final one used in the Z-Wave network model, and it is the "highest" abstraction, the application layer. It is in this layer that the user inputs Z-Wave commands either through direct action such as a remote control or via program or script code. This layer is the user interface (UI), and it may be implemented in hardware, software, or a combination of both. Table 6-1 details the five different packet types used in a Z-Wave mesh network.

Table 6-1   Z-Wave Network Packet Types

Packet Type	Description	Payload Length	Remarks
Multicast	Broadcast to all network nodes	64 bytes max.	0xff address
Singlecast	Sent to a specific node	64 bytes max.	Up to 232 nodes
Routed	Repeated packet (digipeated)	64 bytes max.	4 hops max.
ACK	Acknowledgment	0	—
NACK	Negative acknowledgment	0	—

Every Z-Wave network uses a 4-byte ID called the *Home ID*. Each primary controller has this Home ID, which slave nodes acquire when they join the network. Every secondary controller also uses the same Home ID when they are attached to the network. Individual slave nodes have a 1-byte ID that is assigned by the primary controller when that slave node joins the network.

Network Devices

Z-Wave network *controllers* and *slaves* are the two device types that comprise a Z-Wave network. Slaves are also known as *end point devices* because they only respond to the command packets sent by the controller. Slaves usually contain a microcontroller with GPIO pins that, in turn, control TRIAC (triode for alternating current) devices, which turn AC mains on and off. These GPIO pins are most likely to use optoisolators for maximum protection against inadvertent faults with the AC mains.

Figure 6-4 shows a Z-Wave-enabled duplex outlet I used for the next demonstration. It resembles an ordinary US duplex outlet except for the small push button between the sockets. Of course, the Z-Wave stamping on the front

Figure 6-4   Z-Wave-enabled duplex outlet.

reveals the true nature of this device. This outlet is also about thirty times more expensive than an ordinary "dumb" duplex outlet, but you wouldn't need to have all that many smart control points in your house.

The user must press the button on the duplex outlet to join the Z-Wave network when prompted by the controller device. I will demonstrate how to join the outlet shortly in a demonstration, but I next need to mention a few things regarding the network controllers.

Z-Wave network controllers are either portable or static. Portable controls resemble ordinary AV remote controls, as can be seen in Figure 6-5, which is a GE portable Z-Wave controller. These controllers must be able to self-discover their location within the network topology because they do not have the advantage

of being placed in a fixed location. Such self-discovery is made possible by having the portable controller *ping* nearby nodes that are within their RF range. The controller can then join the network based on the results of the ping packet search. Portable controllers are battery powered and are often used as the primary controller within a Z-Wave network. The controller shown in the figure runs on three AAA batteries, which last a long time.

Static controllers are the other type, and they are so named because they are powered by AC mains and are situated in fixed locations. A static controller can easily monitor all the network traffic and often serves as a secondary controller in an advanced network configuration. The current network configuration may be stored in it, and if so, it is known as a *static update controller* (SUC).

Most often a static controller serves as a bridge between Z-Wave components and devices and non-Z-Wave devices, such as X-10 devices. The static controller acts as a virtual node when deployed in this manner, bridging and converting "foreign" data packets between the outside devices and the Z-Wave network.

A Z-Wave network may have up to 125 virtual nodes, which could be a great help in the case of a large HA system with many legacy devices and components that need to be converted to a Z-Wave network. Static controllers also may serve as TCP/IP gateways, thus allowing Z-wave network Internet connectivity.

Static controllers may act as primary controllers in cases where the normal primary controllers can be placed into a static control proxy situation. In such a setup, this advanced configuration is known as a *SUC ID server* (IS). I will not be using such a complex configuration in any book project, but it does suggest that a large variety of system configurations can be created using Z-Wave controllers.

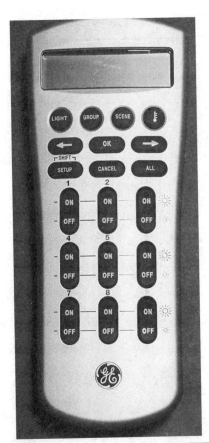

Figure 6-5   GE portable Z-Wave controller.

The Z-Wave Microcontroller

The original Z-Wave chip was designed and manufactured by Zensys, now known as Sigma Designs. All certified Z-Wave component manufacturers must use this authentic Z-Wave chip in their devices. This ensures that any Z-Wave node properly joins the network and communicates with other nodes produced by other manufacturers. The Zensys chip design is discussed in this section because it forms the basis for the complete Z-Wave concept and provides a background for understanding how the RasPi can function as a controller in a network.

A recent Zensys single module is Model ZW3102N, containing a ZW0301 chip that uses the venerable 8051 core with a 32-MHz external crystal. This is a hybrid module containing a lot of additional components, including a RF transceiver operating on either the U.S. or European ISM frequency. There is also a built-in digital modem along with a hardware implementation of the network stack operations that were discussed in the preceding section. The ZW0301 chip has only 32 kilobytes (kB) of Flash memory and a meager 2 kB of static random access memory (SRAM). It operates on a supply voltage range of 2.1–3.6 V DC and consumes a maximum of 36 milliamperes (mA) when transmitting. Figure 6-6 is the block diagram of the ZW3102N showing all the components that constitute this module.

The ZW0301 chip has several of the standard functions that have been discussed in previous chapters, including the serial peripheral interface (SPI) and the universal asynchronous receiver/transmitter (UART) interfaces. The chip also has

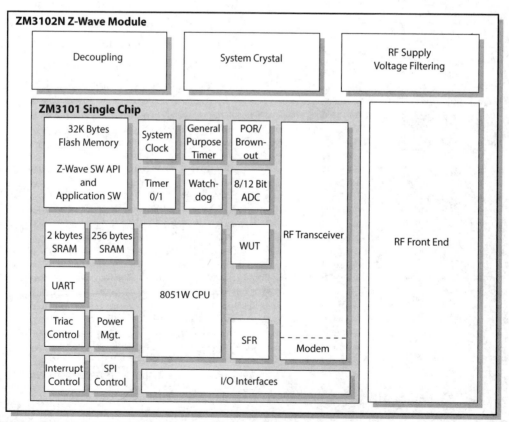

Figure 6-6   Block diagram of the ZW3102N module.

Figure 6-7   ZW3102N module.

timers, interrupts, a watchdog monitor, power management, and brownout detection. It has a four-channel, 12-bit analog-digital converter (ADC), a pulsewidth-modulation controller, and an enhanced TRIAC control with zero crossing detection. A total of 10 GPIO lines are available, but some are multiplexed or shared with other I/O functions. The ZW3102N module is very small. Figure 6-7 shows the module with a U.S. quarter coin for comparison.

The module does need an external antenna, and a few capacitors and inductors complete a Z-Wave device installation. The software is also fixed in the Flash memory and is not available for examination or modification. This where this chapter's RasPi project will open up the Z-Wave network so that you have a chance to experiment with various configurations and monitor network traffic. But first I would like to demonstrate a simple Z-Wave network.

Z-Wave Demonstration

This demonstration uses a portable controller (shown in Figure 6-5) along with two nodes. One node is the duplex outlet shown in Figure 6-4, and the other is an outdoor module, as shown in Figure 6-8.

Please notice the black button located on the top of the device in Figure 6-8. The user needs

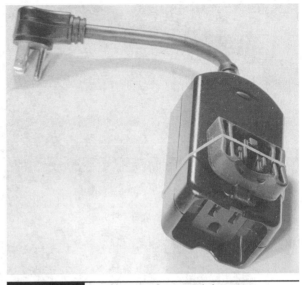

Figure 6-8   Z-Wave outdoor module.

to press this button to join the device to the network when prompted by the controller menu.

The demonstration network nodes or slaves will be made up of the duplex outlet and outdoor module, each controlling a small table lamp. The duplex outlet will actually be connected to a power cord plugged into a regular outlet for this temporary test arrangement. Figure 6-9 shows the test setup on my dining room table.

At first, I arbitrarily assigned device number 4 to the outdoor module and device number 8 to the duplex outlet. I then proceeded to turn the lamp on and off using the portable controller, and everything worked as expected. I was also able to control both devices simultaneously by selecting the "All" mode on the portable controller. The next part of the test was a bit harder because I have a smaller home with an open-plan layout, meaning that there are fewer interior walls than an average cape-style home. I was finally able to place the outdoor module device in the basement and the duplex outlet on the first floor, and I operated the controller in a second-floor bedroom. I was not able to

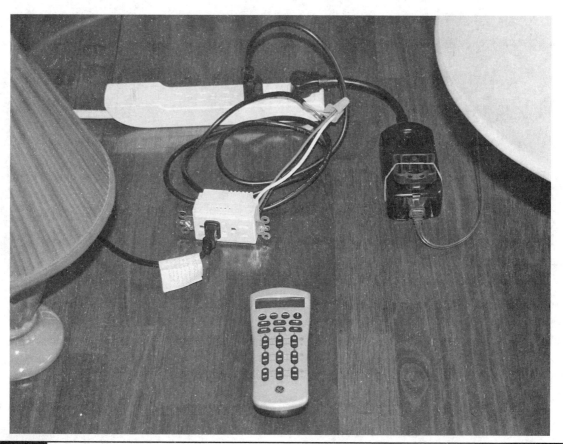

Figure 6-9   Z-Wave test system.

turn on the basement module without having the first-floor module plugged in. This proved that the first-floor module was digipeating and forwarding the control packets to the module located in the basement. The controller showed "Failure" on its liquid-crystal display (LCD) screen with the first-floor module unplugged. This status indicates that no ACKs were being received. Obviously, no NACKs could be sent because the first-floor module was unpowered and the basement module was out of range.

Setting up this demonstration was very simple, and it amply shows that the high-tech Z-Wave network functioned well while providing the user with a very easy-to-use and useful interface. The next demonstration shows how to interface a RasPi with a Z-Wave network. This setup will allow for some interesting experiments.

RasPi and the Z-Wave Interface

Connecting a RasPi to a Z-Wave network requires the use of a Z-Wave USB dongle. One such device made by Aeon Labs, called the *Z-Stick*, is shown in Figure 6-10. It incorporates a Zensys module and a USB interface chip along with some additional firmware to make the two components work together. It also has an internal rechargeable battery that enables the storage of firmware updates and configuration data. The Z-Stick has three operating modes that you should know about:

■ **Inclusion.** This mode adds or includes Z-Wave devices into the network. To add a device:

1. Unplug the Z-Stick from the USB connector.

2. Press the large button on the Z-Stick. The Z-Stick LED will start to slowly blink.

3. Go to the device that you wish to add while holding the Z-Stick, and press and release the device's button.

4. The Z-Stick LED will blink rapidly for several seconds, then glow steadily for 3 seconds, and finally return to a slow blinking state. The device has been added to the network.

- **Removal.** This mode will remove or exclude Z-Wave devices from the network. To remove a device:

  1. Unplug the Z-Stick from the USB connector.

  2. Press and hold the large button on the Z-Stick for about 3 seconds. The Z-Stick LED will start to blink slowly and then transition to a fast blink.

  3. Go to the device that you wish to remove while holding the Z-Stick, and press and release the device's button.

  4. The Z-Stick LED will then glow steadily for 3 seconds and finally return to a fast blinking state. The device has been removed from the network.

- **Serial API.** This is the mode where the Z-Stick acts as the portal between the RasPi and the Z-Wave network. Simply plug it into a powered hub USB connector because the RasPi does not have sufficient power for the Z-Stick. RasPi software will now take control of the Z-Wave network.

I now have to take a brief detour from the Z-Wave to introduce the Secure Shell login process that will be used in establishing the control software environment.

Figure 6-10 Aeon Labs Z-Stick.

SSH Login

In this section, I will show you how to log onto the RasPi using a network connection. The Stretch Linux distribution, as well as many others, includes a great service known as *Secure Shell* (SSH). It is a network protocol that uses cryptographic means to establish secure data communication between two networked computers connected via a logical secure channel over a physically insecure network. SSH uses both server and client programs to accomplish the connection.

One of the questions that arises when you are first configuring your RasPi is whether or not to start SSH on boot-up. I recommended that you answer yes because that automatically starts the SSH daemon each time you start the RasPi. The second part of the connection is the client program, which is highly dependent on the type of computer you are using to connect to the RasPi. I recommend using putty.exe because most of you will be using a Windows-based machine. Putty is freely available from a variety of Internet sources, so I would recommend a Google search to locate a good download mirror. You should see the Figure 6-11 screenshot, assuming that you answered yes to the SSH question and have downloaded and are running Putty on a Windows-based computer

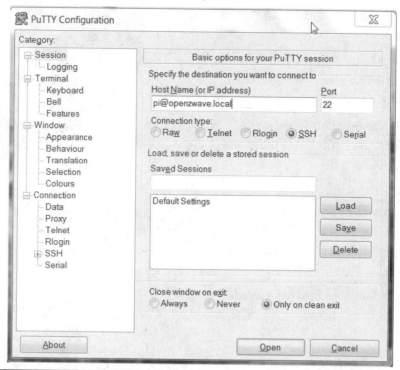

Figure 6-11   Putty screenshot.

connected to the same network that connects to the RasPi.

Don't be concerned with the host name that appears in the screenshot; I will get to that shortly. When you click on the Open button at the bottom of the Putty screen, you will see a screenshot of a RasPi terminal window asking, in this case, for a login password (Figure 6-12).

At this point, you are in a RasPi terminal window session—absolutely no different from looking at a monitor connected directly to a RasPi and using a locally connected keyboard and mouse. This transparency is what makes SSH so great; it allows you to remotely log in to the RasPi without being concerned with any minutiae about the connection. You may type

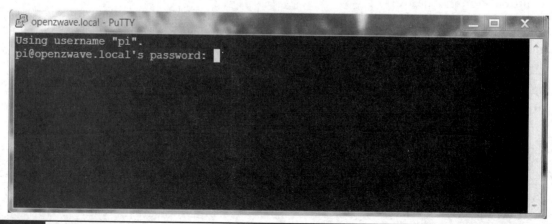

Figure 6-12   RasPi terminal window.

in any normal command and have the RasPi respond as appropriate.

I will now return to my Z-Wave software discussion, now that you are familiar with SSH.

Project Things by Mozilla.org

Project Things is a software framework consisting of applications, utilities, and services that can connect smart HA devices to computers and microcontrollers such as the RasPi. The latest implementation of Project Things is called the *Things Gateway*. It will allow you to directly control and/or monitor HA devices and appliances over the Web. It can replace all the separate mobile apps that can quickly accumulate for individual HA devices, systems, and appliances. Mozilla has made available a full RasPi Debian Linux distribution that contains the Things Gateway preinstalled and ready for deployment. You will need to use an OpenZWave-compatible dongle (adapter) to be able to communicate with any existing Z-Wave devices. I used an older Aeon Labs Z-Stick S2 dongle that I had from use in previous projects. You can also use this if it still available; otherwise, use either one of the following:

- Sigma Designs UZB Stick

- Aeotec Z-Stick (Gen5)

Ensure that the stick uses the correct frequency for your region, which I discussed in an earlier section.

Software Installation

Much of the following discussion is based on a blog tutorial written by Ben Francis that is available at https://hacks.mozilla.org/2018/02/how-to-build-your-own-private-smart-home-with-a-raspberry-pi-and-mozillas-things-gateway/. The first thing you will need to do is to download the disk image from https://iot.mozilla.org/gateway. The current gateway was version 0.4 at the time of this writing.

Next, create a bootable micro SD card using the procedures discussed in Chapter 1. Mozilla also has instructions on how to run the gateway software on a PC or Mac if you are inclined to experiment. However, I will be staying with the RasPi installation for this book.

The gateway software will start a WiFi hot spot with the SSID name of *Mozilla IoT Gateway*. You should join this network in order to connect the RasPi to your own WiFi network. If you are using another computer on your network or a smartphone, the Mozilla IoT Gateway will show all the local WiFi networks detected, as shown in the Figure 6-13 tutorial example. Select your regular network, and enter the passkey or pass phrase for a secured network.

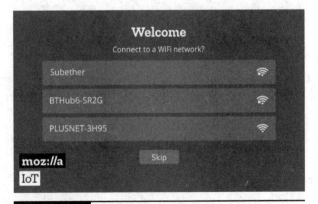

Figure 6-13 WiFi network selection.

Figure 6-14   Connecting to the local WiFi network.

Figure 6-16   Account setup screen.

If you are using the RasPi browser, go to http://gateway.local/ and start the setup process. Figure 6-14 shows the screen you will see while the gateway connects to your WiFi network.

You will be asked to enter a unique subdomain name once you are connected to the gateway URL. You will also need to enter an e-mail address, which will enable you to retrieve your subdomain name at any future time. The subdomain name is used, in part, to generate an Secure Sockets Layer (SSL) certificate, which, in turn, is used to establish a secure Internet tunnel so that you can remotely access the gateway. Figure 6-15 shows the screen for entering these data.

After setting the secure gateway address, you will be placed into your new subdomain, and another dialog box will appear, where you will enter data to create an account in the gateway. This data consist of your name, e-mail address, and password. Figure 6-16 shows this account setup screen.

This last step completes the gateway configuration/setup, and you will now see an Add Things screen, as shown in Figure 6-17. How to add things is the subject for the next section.

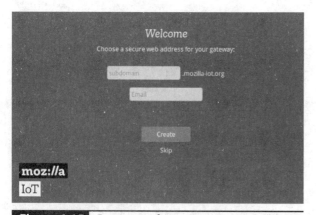

Figure 6-15   Data entry for secure gateway address.

Figure 6-17   Add Things screen.

Add Things

You add things to the gateway by clicking on the "+" icon located at the lower right-hand corner, as shown in Figure 6-17. I did this, and Figure 6-18 shows the things or devices that were discovered.

Any discovered device must be linked or paired subsequently with the gateway. This is easily done with Z-Wave devices by following the procedure I detailed earlier. When the pairing is completed, click on the Done button shown

in the figure, and the newly paired devices will appear on the Things screen, as shown in Figure 6-19.

Test Run

The Z-Wave test of the gateway system was rather simple. I simply plugged an AC lamp into the outdoor Z-Wave device and turned the lamp on and off by clicking on the Switch icon shown in the figure. Incidentally, I am unsure what the Z-Wave-16a124a-4-Sensor icon shown in the figure refers to. It may be an undocumented feature for the switching device. In any event, clicking on it did not cause an action to be observed regarding the AC lamp plugged into the switch.

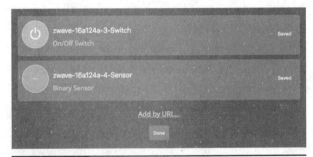

Figure 6-18 Discovered Z-Wave devices.

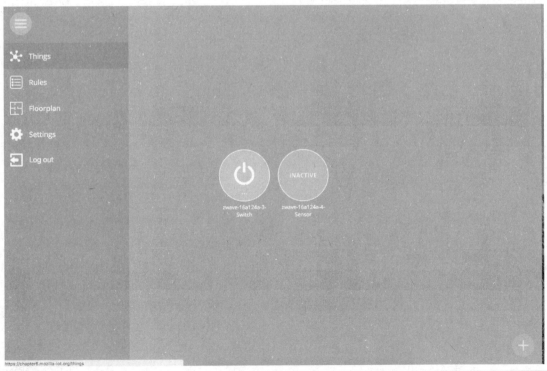

Figure 6-19 Things screen.

Summary

I started this chapter with a definition of home automation and provided a list of representative tasks that would fall under the broad umbrella that comprises home automation.

Next was a discussion of the underlying base technologies that support HA implementations. I provided a rationale for selecting the Z-Wave protocol because it is a modern, highly flexible system that is very easy to configure and lends itself quite well to a simple RasPi interface connection.

Two representative Z-Wave devices were described next, along with a working small-scale Z-Wave network. A network demonstration was created using a commercial portable controller device.

The core Z-Wave chip was discussed to provide you with a better understanding of how the overall network functions and how devices are highly dependent on the chip functions to be added seamlessly to the network. A Z-Stick was shown that enables a RasPi to Z-Wave interface to be implemented. Several alternatives to the Z-Stick were also discussed

I next showed you a highly useful SSH remote login procedure that you can use to access the RasPi while the open-source Z-Wave software is running. The open-source software executes as a Web service, thus making the RasPi GUI inaccessible. Any user access to the RasPi must then be done using a remote access service such as SSH.

I used the Project Things framework by Mozilla to create a gateway to control the Z-Wave devices introduced earlier in the chapter. This project's version generates a gateway that you use to access and control Z-Wave devices. The Z-Stick is used to communicate between the RasPi and any nearby Z-Wave devices. I went through a detailed discussion of how to install and configure the gateway. The chapter ended with a simple demonstration of turning an AC lamp on and off using commands sent through the gateway.

## CHAPTER 7

Mycroft and Picroft

THIS CHAPTER INTRODUCES an open-source personal voice assistant project named *Mycroft*. Mycroft has a RasPi variant named *Picroft*, which I will discuss after going through the Mycroft introduction.

Mycroft's home website is mycroft.ai. Mycroft is the invention of Ryan Sipes and Joshua Montgomery, who liked the idea of creating a simple and basic intelligent virtual assistant, similar to commercial ones in existence at the time but based entirely on an open-source concept. Mycroft uses only open-source components for the intelligent personal assistant and knowledge navigator and uses the Linux OS as a platform. The Mycroft developers were very much concerned with privacy issues and strongly believed that developing an open-source solution would strongly address those concerns. It turns out that they were entirely correct in this area because privacy concerns have recently arisen concerning both the Google and Amazon systems. No such problems are anticipated with the Mycroft solution.

The Mycroft team has raised funds through a variety of sources, including private equity investors, and has offered shares of the company to the public through Startengine, an equity crowdfunding platform. The project itself is not

Parts List

Item	Model	Quantity	Source
RasPi 3	B	1	adafruit.com amazon.com mcmelectronics.com
USB microphone	Fifine USB Microphone	1	amazon.com
USB speakers	ARVICKA Blue LED USB Speakers	1	amazon.com
Tactile push button	Commodity	1	adafruit.com
2.2-kiloohm (KΩ), ¼-watt (W) resistor	Commodity	1	adafruit.com
Philips Hue smart lamp	White	1	Home improvement store amazon.com
Philips Hue bridge	Philips Hue Smart Hub	1	amazon.com
AC table lamp	Commodity	1	Home improvement store

named after Sherlock Holmes's older brother but instead after a fictional computer from Robert Heinlein's 1966 science fiction novel, *The Moon Is a Harsh Mistress*, which I strongly recommend that you read if you are sci-fi fan.

Mycroft Structure

Mycroft is designed to run on many different platforms, which are called *Devices* in Mycroft terminology. Different hardware implementations that host Mycroft are called *Enclosures*. The currently supported Enclosures are

- **Mark 1.** A software image of Mycroft designed to be installed on the Mycroft Mark 1, a reference hardware device described in a later section

- **Picroft.** A software image of Mycroft designed to be installed on the RasPi 3, Model B

- **Android.** A software image of Mycroft designed to be installed on Android devices

Mycroft uses an open-source intent parser called *Adapt* that converts natural language into machine-readable data structures. An intent parser is just a library for converting natural language into machine-readable data structures, such as JavaScript Object Notation (JSON). It is lightweight and has been designed to run on devices with limited computing resources, such as embedded devices including the RasPi. Adapt takes in natural language as an input and outputs a data structure that includes

- **Intent.** What the user is trying to do

- **Probability match.** A measure of how confident Adapt is that the intent has been correctly identified

- **Tagged list of entities.** Objects that are used by Mycroft Skills to perform actions or functions

Adapt is important for interpreting the user's natural-language input. For example, you might want to create a voice user interface that allows a user to play a Pandora station. The utterances a user might say include

- "Turn on Pandora."

- "Play Pandora."

- "Play my Jimmy Buffett Pandora station."

The Adapt intent parser takes this input and generates a JSON data structure similar to this:

```
{

    "confidence": 0.61,
    "target": null,
    "Artist": "jimmy buffett",
    "intent_type": "MusicIntent",
    "MusicVerb": "put on",
    "MusicKeyword": "pandora"

}
```

Applications, which are Mycroft Skills in this case, can then parse the JSON data and take appropriate action, such as playing Jimmy Buffett using the open-source Pandora application. I will discuss Mycroft Skills in much greater detail in a later section.

Mycroft uses Mimic for speech synthesis, which, in turn, is based on the Festival Lite speech synthesis system. Festival Lite was developed at the Carnegie Mellon University (CMU), where it is known as *Flite* (Festival Lite). It is a small, fast run-time open-source text-to-speech synthesis (TTS) engine designed primarily for small embedded machines and/or large servers. Again, this TTS library is targeted to be used within embedded platforms such as the RasPi.

Mycroft was always designed to be modular, enabling users to change its components. For example, the eSpeakNG TTS library can be substituted for Mimic if a user finds it more appropriate.

The *wake word* is the word or phrase that a user speaks to alert Mycroft that an utterance will follow. Mycroft uses the PocketSphinx technology developed by CMU for wake-word detection. PocketSphinx is a lightweight version of CMU's larger speech-recognition package named *CMUSphinx*. PocketSphinx is well suited for embedded platforms just the way the other Mycroft packages have been designed.

You may change the wake word from the default "Hey mycroft" to anything phrase you want by going to the home account website at https://home.mycroft.ai/ and entering your own phrase. There is a constraint on wake-word selection that it must be in the English language. Wake words in German, French, Spanish, and so on will not work. However, the Mycroft development team is currently working on an open-source software package known as *Precise* that will recognize any wake word in any language. This is because Precise uses an artificial neural network (ANN), which can be trained to recognize any series of audio sounds. Of course, this means that you must train Precise to recognize your chosen wake word before it can be used with Mycroft. It will still be a while before Precise is made freely available for use by the Mycroft open-source development community.

Mycroft, like all the personal voice assistants, has speech-to-text (STT) software, which is used to take any user's spoken words and turn them into text phrases that can then be further processed. Mycroft currently uses Google's STT, but this selection may be changed due to Mycroft's modular design. IBM's Watson STT can be substituted as well as Facebook's wit.ai SST package. Mycroft is planning on developing its own open-source STT solution, which will be named *OpenSTT*.

Mycroft uses middleware, which is software that is akin to an HA OS. This middleware has two components:

- **Mycroft Core.** This is Python code, which is the core software that "glues" all the other Mycroft modules together. The Mycroft Core code is available from GitHub under its open-source license.

- **Mycroft Home and Mycroft API.** This is the software infrastructure that holds all user and device data. This infrastructure provides high-level services, including storing API keys, which are used to access third-party services that provide Skill functionalities. This code is considered proprietary by Mycroft and has not been made available as open source. This decision does not violate the open-source license that Mycroft uses to make the majority of its software freely available.

I will not be discussing any further Mycroft structural changes, but I will alerting you to the fact that it can be done. The Mycroft software was released using the Apache Foundation open-source library. The important point to be made is that both the Google and Amazon systems are completely closed source, and users are not able to make these types of changes to suit their own situations.

Mycroft Hardware

The Mycroft project is not solely about open-source software. The company does offer hardware that runs its software. The first-generation hardware released by the company was the Mycroft Mark I shown in Figure 7-1. Incidentally, this hardware is still available for purchase, with the first units shipping in April 2016. All the company's hardware is open source, released under the CERN open hardware license.

Figure 7-1   Mycroft Mark I.

Some key technical specifications for the Mark I unit are as follows:

- Raspberry Pi 3, Model B
- Built-in speaker
- RCA audio output ports
- 8 × 32 LED display
- Dual NeoPixel "eyes"
- Built-in WiFi (802.11B/G/N)
- 10/100 Ethernet port, HDMI debug port, 4 USB ports, 40-pin GPIO connector
- Integrated Arduino Mini
- Enclosure with side-mount ports
- U.S. power adapter

NOTE: The 40-pin connector has a different pinout than an unmodified RasPi. The RasPi's built-in Bluetooth is also disabled.

Figure 7-2 shows the back of the Mark I, where you can clearly see the normal layout

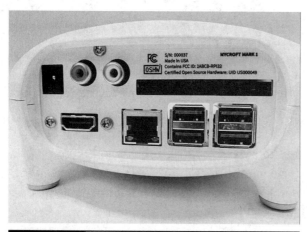

Figure 7-2   Mark I back panel.

of four USB ports as well as the RJ45 socket belonging to the enclosed RasPi. The two RCA phone jacks shown in the figure are for external stereo audio, in case the internal speaker is insufficient. There is also an HDMI socket for an external monitor as well as a DC power socket for an AC wall wart supply, which comes with the unit.

Figure 7-3 is another view of the unit that shows a large push button protruding from the case top. The push button supports multiple functions, some of which are discussed in a later section detailing Mycroft Skills. In any case, a detailed operational PDF guide on the Mark I is available from the Mycroft website.

The company announced the Mark II hardware project in early 2018. Figure 7-4 is a photograph from the company's Kickstarter campaign. This next-generation hardware will

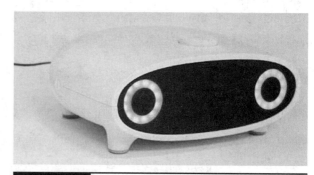

Figure 7-3   Oblique view of the Mark I.

Figure 7-4   Mark II prototype.

include a visual display, making it somewhat akin to the Amazon Show unit. I am expecting that this unit will likely be available in late 2018 or early 2019.

The company just announced another unit, the Mark III, to be released in 2019. No further details were available at the time of this writing regarding the Mark III.

The preceding sections all concerned the original Mycroft software and associated hardware that runs that software. The next sections discuss Picroft, which is Mycroft designed for the RasPi.

Picroft

Picroft is the Mycroft software package designed to function with an unmodified RasPi 3, Model B. This is a logical extension for Mycroft because the Mark I uses a RasPi 3, Model B, as its main processor. I will next detail how to install Picroft, configure it, and finally test it.

Picroft Installation

A complete set of Picroft installation instructions is available at the Picroft website (https://mycroft.ai/documentation/picroft/). These instructions also include a list of all the hardware needed for a RasPi Picroft installation. These items are a RasPi 3, Model B, a USB microphone and speaker(s), and the usual desktop configuration accessories.

There are two ways to install Picroft on a RasPi 3, Model B. The first way is the one I describe in detail in this section and the one that I highly recommend that you follow. However, you can install and build Mycroft from scratch using the instructions found at the website https://mycroft.ai/documentation/linux/-installing-via-git-clone. This installation takes several hours and must be done using a clean Jessie Raspbian disk image. I would only recommend the second approach for experienced Linux developers who are confident in their knowledge and ability to resolve any software dependency issues.

The simplest way to install the Picroft software is to load it in the form of a disk image that is stored on a bootable micro SD card. The instructions provided for writing a micro SD card in Chapter 1 also apply to this installation. The following steps should be done in the sequence presented:

1. Boot the RasPi after you have burned the Picroft disk image.

2. Connect to the WiFi SSID MYCROFT using another computer on the same network that is connected to the RasPi.

3. In a browser, go to the Web page http://start.mycroft.ai. A list of WiFi networks will be displayed. Select the WiFi network that you want to connect to the Picroft, and enter the passkey or passphrase for that network.

    NOTE: Picroft cannot connect to WiFi networks using the 5-GHz band. This usually means that any network with a "5G" in its name is not to be used. In addition, 2.4-GHz WiFi networks using channels 12 or 13 (2.467 and 2.472 GHz) cannot be used.

4. The Picroft software will proceed to automatically connect to the Internet and download a considerable amount of data. Please be patient while this download takes place. Once the download has finished, a message will be displayed on the monitor providing a registration website and a five- or six-alphanumeric-character sequence that you need to enter at the website. The registration website that I used was https://home.mycroft.ai/device. Yours may be different because these installation procedures do change occasionally. The

registration character sequence will also be repeatedly spoken through the USB speaker. Mycroft refers to this process as *pairing*, not to be confused with Bluetooth pairing.

5. The next step is to set up an SSH link with the RasPi. There are several ways to accomplish this task:

    a. Go to your WiFi router's admin login, and examine the attached wireless devices and try to determine your RasPi's IP address. Use this IP address to connect to the RasPi per the SSH instructions provided in Chapter 6. Please note that the password for SSH login is "mycroft" not the usual "raspberry."

    b. Go to the Mycroft documentation website https://mycroft.ai/ documentation, and read about how to add the IP address. Then speak the phrase "Hey Mycroft, what's your IP address?" The response both spoken and on the screen should be "Here are my available IP addresses: wlan IP address … Those are all my available IP addresses." A word of caution is in order here; this procedure may not work because your USB microphone may not be enabled. I will shortly discuss how to enable the USB microphone.

    c. Press CTRL-C to exit Picroft and get to a terminal prompt on the RasPi monitor. Then enter the command `ifconfig`, and the current IP address will be shown in the `wlan0` section. You will have to reboot the RasPi to restart Picroft.

At this point, you should have fully functional Picroft system. However, I found that my USB microphone still was not working properly with the software. This situation required me to make a configuration change to the pulse audio system that Picroft uses for its audio sinks and sources.

The procedure I used for this configuration change follows:

1. Find the proper USB identification for the microphone. One way is to enter the following command at a terminal prompt:

```
pactl list short sources
```

2. Edit the pulse audio file named *default.pa*. It is in the directory /etc/pulse. My edit consisted of changing one of the last two lines in the file from `set-default-source input` to `set-default-source alsa_input.usb-Samson_Technologies_Samson_C01U-00-C01U.analog-stereo`. The last portion of the line came from the `pactl` command

3. Reboot the RasPi to allow the configuration change to take effect.

I found that the Picroft functioned as expected after I enabled the USB microphone. I tested the system by speaking this phrase: "Hey Mycroft, what time is it?" The response was both spoken and displayed on the Picroft log screen, as shown in Figure 7-5.

Configuring Mycroft

I would like you to perform a simple experiment before I explain how to configure the Mycroft software. Ask Mycroft the following: "Hey Mycroft, what is my location?" The answer will likely startle you, as it did me. The response both in text and spoken was: "I'm in Fredonia Kansas United States."

This is a very strange response at first glance until you realize that the location is hardcoded into the Mycroft configuration file and also takes location data from the Mycroft home website (https://home.mycroft.ai/). The basic Mycroft configuration file is named *mycroft.conf* and is stored at

Figure 7-5 Picroft log screen for the utterance "Hey Mycroft, what time is it?"

```
/opt/venvs/mycroft-core/lib/python3.4/
    sitepackages/mycroft/configuration/
    mycroft.conf
```

And yes, that location has eight directory levels to it, not the easiest to find. The configuration file has four sections, which are loaded in the following progression:

■ DEFAULT

■ REMOTE

■ SYSTEM

■ USER

Because the USER level is the last to be loaded, it can always override any settings or configurations loaded by any one of the prior levels. The following listing was extracted from the DEFAULT section of the configuration file. In this listing, you can clearly see where the Kansas and United States locations came from.

The Fredonia word came from the Location textbox entry from the home website, as shown in Figure 7-6. This entry became an override for the Lawrence name shown in the basic configuration file.

Figure 7-6 Location entry from the Device page in the home account.

```
// Location where the system resides
// NOTE: Although this is set here, an
// Enclosure can override the value
// For example a mycroft-core running in
// a car could use the GPS.
// Override: REMOTE
"location": {
  "city": {
    "code": "Lawrence",
    "name": "Lawrence",
    "state": {
      "code": "KS",
      "name": "Kansas",
      "country": {
        "code": "US",
        "name": "United States"
      }
    }
  },
  "coordinate": {
    "latitude": 38.971669,
    "longitude": -95.23525
  },
  "timezone": {
    "code": "America/Chicago",
    "name": "Central Standard Time",
    "dstOffset": 3600000,
    "offset": -21600000
  }
}
```

Mycroft GPIO Skill

Mycroft Skills are add-ins or plug-ins that provide additional functionality to the Mycroft software package. Skills have been developed by both Mycroft company developers and open-source Mycroft community developers and vary greatly in their functionality and maturity. Hopefully, you will be able to develop your own custom Skill after becoming familiar with Mycroft development and experimenting with this particular Skill. I chose the GPIO Skill because it adds an important function to Mycroft, allowing a user to directly control RasPi GPIO pins in support of an HA

application. You will be able to both read from a GPIO port by detecting a pushbutton press and write to it by lighting a LED.

Follow this sequence of steps to download and install the GPIO skill:

1. Use a computer on the local network and go to the GitHub webpage at https://github.com/MycroftAI/picroft_example_skill_gpio. Download and extract the zip file from this page. This will create a directory named *picroft_example_skill_gpio_master* on the computer.

2. Make an SSH connection to the RasPi loaded with the Picroft software.

3. Transfer the new directory to the RasPi. If you are using a Mac or Linux terminal window, then the command is similar to this one that I used:

    ```
    scp -r ./Desktop/picroft_example_skill_
    gpio-master pi@192.168.1.29:picroft_
    example_skill_gpio-master
    ```

4. Stop the Mycroft server if it is running on the RasPi, which will cause a terminal prompt to be displayed.

5. Enter the following commands:

    ```
    cd picroft_example_skill_gpio_master
    sudo nano Makefile
    ```

    ■   Edit the `Makefile` so that the IP address displayed everywhere in the editor is the RasPi IP address.

    ■   Press CTRL-O and CTRL-X to save and exit the editor.

6. `cd /opt/mycroft/skills`

7. `mkdir skill-gpio`

8. `cd ~/picroft_example_skill_gpio_master`

9. `make install.pi`

10. `cd /opt/mycroft/skills/`

11. `sudo chown mycroft:mycroft -R skill-gpio`

12. `sudo adduser pi mycroft`

13. `sudo adduser mycroft gpio`

At this point, the GPIO Skill should be ready for a test. However, you will now need to set up the test circuit.

GPIO Test Circuit

Figure 7-7 is a Fritzing diagram that shows the test circuit to be used for the initial GPIO Skill demonstration.

Test Run

The test run involves two phases. The first one is the direct programmatic control of the hardware attached to the RasPi GPIO pins. This will be accomplished by issuing the following commends either at the RasPi terminal prompt or using an SSH session:

```
cd ~/picroft_example_skill_gpio-master
make test.pi
```

I used an SSH session to build the test script `test.pi`. After the build process finished, the LED slowly started blinking for approximately 10 seconds. You should also press the push button sometime within that same 10-second interval. Figure 7-8 shows the output generated by the `GPIO.py` Python script that runs during this test.

The script shows the continuous LED status while the script ran. The LED logical name in the script is `GPIO1`, and the status is either `On` or `Off`. The push button logical name is `Button`, and its status is either `Pressed` or `Released`. You will probably notice that there are multiple occurrences of `Pressed` and `Released` for the push button operations. In reality, I only pressed and released the button one time. The multiple recorded button states are due to the fast polling taking place in the script.

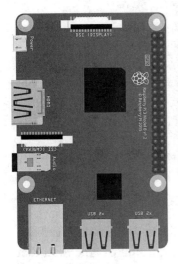
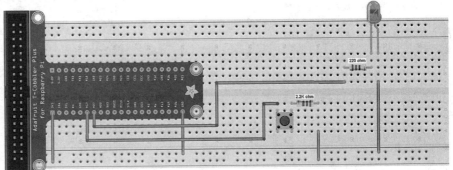

Figure 7-7   Fritzing diagram for GPIO test circuit.

```
donnorris — pi@picroft: ~/picroft_example_skill_gpio-master — ssh pi@192.16...
pi@picroft:~/picroft_example_skill_gpio-master $ make test.pi
ssh pi@192.168.1.29 python /opt/mycroft/skills/skill-gpio/GPIO.py
pi@192.168.1.29's password:
{'GPIO1': 'On'}
{'GPIO1': 'Off'}
{'GPIO1': 'On'}
{'GPIO1': 'Off'}
{'GPIO1': 'On'}
{'GPIO1': 'Off'}
{'GPIO1': 'On'}
{'GPIO1': 'Off'}
{'GPIO1': 'On'}
{'GPIO1': 'Off'}
{'GPIO1': 'On'}
{'GPIO1': 'Off'}
{'GPIO1': 'On'}
{'GPIO1': 'Off'}
{'GPIO1': 'On'}
{'GPIO1': 'Off'}
{'Button': 'Pressed', 'GPIO1': 'On'}
{'Button': 'Pressed', 'GPIO1': 'Off'}
{'Button': 'Released', 'GPIO1': 'On'}
{'Button': 'Released', 'GPIO1': 'Off'}
{'Button': 'Released', 'GPIO1': 'On'}
GPIO is valid
{'Button': 'Released', 'GPIO1': 'Off'}
pi@picroft:~/picroft_example_skill_gpio-master $
```

Figure 7-8   GPIO test diagram.

The second phase for testing involves using voice control with the Mycroft service operating. Speak the following phrases to test GPIO control:

- "Hey Mycroft, turn LED on."
- "Hey Mycroft, turn LED off."
- "Hey Mycroft, blink LED."

The system will respond by turning the LED on, then off, and finally, blinking it. It also will speak the responses, such as

- "The LED is on."
- "The LED is off."

These responses are continually repeated for the blink operation. In fact, the only way I could stop the blinking was to use a keyboard interrupt (CTRL-C) to stop the Mycroft server.

An additional 26 skills are provided with the disk image in addition to the GPIO Skill just discussed. I provide a list of these Skills and some further explanations of the more useful or interesting ones in the next section.

Mycroft Skills

Table 7-1 details all the available Skills in the downloaded disk image included with the GPIO Skill that was discussed in the preceding section. Most of the table description content is based on the README.md file found in every Skill directory.

Philips Hue Skill

It is reasonably easy to add an HA Skill to the Picroft system. I selected a Philips Hue Skill that can turn on and off a white Hue light, which I had demonstrated in Chapter 5. This Skill was created by Christopher Rogers and is available

Table 7-1 Available Mycroft Skills

Skill Name	Description
`skill-alarm`	Set daily alarms, recurring alarms, or one-time alarms with Mycroft.
`skill-audio-record`	Audio functions: record, play, stop, and cancel.
`skill-configuration`	Change the technology used to perform wake-word spotting, the system that wakes the device up when you say "Hey Mycroft."
`skill-date-time`	Get the local time or time for major cities around the world. Times are given in 12-hour (2:30 pm) or 24-hour format (14:30) based on the Time Format setting at https://home.mycroft.ai/#/setting/basic.
`skill-gpio`	Demonstrates interacting with the RasPi GPIO pins. Read from a GPIO pin (detecting a button press) and write to one (lighting an LED).
`skill-hello-world`	*Usage:* "Hello world," "How are you?," and "Thank you."
`skill-installer`	Add and remove Skills using the Mycroft Skill Manager (MSM). Install a Skill verbally by saying "Install <skill identifier>," where <skill identifier> is the full name or at least an adequate subset of the name to uniquely identify the Skill.
`skill-ip`	Retrieve the network address (aka Internet Protocol [IP] address) to which the Mycroft device is connected.
`skill-joke`	Brighten your day with a little humor. This draws on the jokes collected by the PyJokes project (https://github.com/pyjokes/pyjokes) to give you a chuckle. The joke categories are: *Neutral:* jokes that are safe for work, kids, or your grandmother. *Adult:* nothing horrible, but be ready to cover some ears. *Chuck Norris:* jokes only a geek can love.
`skill-mark1-demo`	The Mycroft Mark 1 menu, which appears when you press and hold the top button, has a "demo" option. This Skill implements a simple mode that can be used to draw attention at trade shows, stores, etc. The demo starts with the unit's eyes dancing around. Every two minutes it will sing a song. The singing is synched to the clock, so multiple units can form a chorus. You can stop the demo by pressing the top button or saying "Stop."
`skill-naptime`	Tell Mycroft to sleep when you don't want to be disturbed in any way. This stops all calls to the Speech-to-Text system, guaranteeing that your voice won't be sent anywhere on an accidental activation. When sleeping, Mycroft will only listen locally for the phrase "Hey Mycroft, wake up." Otherwise, the system will be totally silent and won't bother you. On a Mark 1, this also dims the LED eyes.
`skill-npr-news`	Plays the latest news from a configurable RSS-based audio feed. By default, the NPR hourly news broadcast is used, but you can choose from other news feeds, including BBC, AP, CBC, CNN, PBS, and Fox. See the setting at https://home.mycroft.ai/#/skill.
`skill-pairing`	The default back-end to provide services for Mycroft users is https://home.mycroft.ai/. Pairing a device with Home provides access to privacy-protecting Speech-to-Text, Wolfram Alpha, and other such services, as well as easy configuration for all your Mycroft devices.

(continued on next page)

Table 7-1 Available Mycroft Skills (*continued*)

Skill Name	Description
skill-personal	This Skill will answer some of the personality questions relating to Mycroft, such as "What are you?," "Where were you born?," and "Who made you?."
skill-playback-control	This Skill doesn't do anything by itself, but it provides an important common language for many audio playback skills. By handling simple utterances such as "Pause," this one Skill can turn around and rebroadcast the message bus command mycroft.audio.service.pause. This lets several music services share the common "Pause" terminology.
skill-reminder	*Usage:* "Remind me to search about AI in 10 minutes."
skill-singing	*Usage:* "Sing." There are five mp3 songs in the skill directory.
skill-speak	Turn Mycroft into a parrot. Speak a phrase and listen to it repeated in Mycroft's dulcet voice! Examples: "Say Goodnight, Gracie"; "Repeat Once upon a midnight dreary, while I pondered, weak and weary, Over many a quaint and curious volume of forgotten lore"; "Speak I can say anything you'd like!"
skill-spelling	*Usage:* "Spell Mycroft."
skill-stock	*Usage:* "Stock price of Google," "trading at Google."
skill-stop	*Usage:* "Stop."
skill-support	Generate a package with debugging information and have it sent to your registered account. You can use this packet to debug issues yourself, or it can be sent on to the support team. This Skill uses the http://termbin.com/ service for storing the debugging information. Examples: "Create a support ticket," "You're not working!," "Send me debug info."
skill-version-checker	Report the version of your Mycroft install (mycroft-core) and of the platform you are running on (e.g., "Mark 1, build 10"). Examples: "Check version," "What version are you running?," "What's your platform build?"
skill-volume	Control the volume of Mycroft with verbal commands or by spinning the physical button on a Mark 1. Examples: "Turn up the volume," "Decrease the audio," "Mute audio," "Set volume to 5," "Set volume to 75 percent."
skill-weather	Get weather conditions, forecasts, expected precipitation, and more! By default, it will tell you about your default location, or you can ask for other cities around the world. Current conditions and weather forecasts come from Open Weather Map at https://openweathermap.org. For devices with screen support, conditions are briefly shown. Examples: "What is the weather?," "What is the forecast tomorrow?," "What is the weather going to be like Tuesday?," "What is the weather in San Francisco?," "When will it rain next?," "How windy is it?," "What's the humidity?"
skill-wiki	Query www.wikipedia.org for answers to all your questions! Get just the summary, or ask for more to get in-depth information. Examples: "Tell me about Elon Musk," "Tell me about beans," "Check Wikipedia for beans," "Search for water."

from GitHub at the website https://github.com/ChristopherRogers1991/mycroft-hue. You can directly clone and install it by entering the following commands, assuming that you have git already installed:

```
cd /opt/mycroft/skills
sudo git clone https://github.com/
    ChristopherRogers1991/mycroft-hue
sudo apt-get install python3-pip
cd mycroft-hue
sudo pip3 install -r requirements.txt
```

The next part of this Skill installation process is to add some information to the basic *mycroft.conf* file concerning the specific Hue device to be controlled. Enter the following to change Mycroft's configuration to recognize the Hue light:

```
cd /opt/venvs/mycroft-core/lib/python3.4/
    sitepackages/mycroft/
configuration/mycroft.conf
sudo nano mycroft.conf
```

Enter this text into the middle of the Skills configuration section:

```
...
"PhilipsHueSkill": {
    "ip": "",
    "username": "",
    "verbose": false,
    "brightness_step": 50,
    "color_temperature_step": 1000,
    "default_group": 0
  },
```

NOTE: If you know the IP address of your hub and/or if you have a username that you would like to use, you may add either or both to the preceding relevant lines. If not, when the Skill is first used, it will attempt to find the hub on your network. If there are multiple hubs, it will take the first one it finds and will create a default user and record the IP address.

On your first run, if you did not supply a username, when you say any phrase that gets routed to this Skill (i.e., "Turn off my lights"), you will be asked to push the button on the top of your Philips Hue hub. This will create a user on the hub for this application.

Save and exit the nano editor, and restart the Mycroft server for the new configuration changes to take effect.

I provided both the Hue bridge IP address and username. The system turned on the Hue light when I spoke the phrase "Hey Mycroft, turn on the workplace light."

This result confirmed that the Mycroft Hue Skill was working correctly and that I had successfully created a simple HA application for the Mycroft system. You can add similar HA Skills following the same procedure I just detailed. Additional HA skills are continually being added to the Mycroft Skills repository by the open-source community developers.

Summary

The chapter began with an introduction to the Mycroft, which is a completely open-source artificial intelligence (AI) project involving a personal voice assistant. I examined its structure and supporting hardware. The Mycroft team offers a hardware device for sale named Mark I that contains a RasPi 3 that hosts the Mycroft software package. The open-source Mycroft software can also be installed on your RasPi 3, in which case it is known as Picroft. Finally, there is a Mycroft version that can be hosted on Android devices.

I reviewed how to install Mycroft on a RasPi. In the process of this installation, I uncovered and resolved an issue in which the software did not discover the USB microphone. I successfully demonstrated that the Picroft installation worked as expected.

The subject of Mycropft Skills was discussed next, and a Mycroft GPIO Skill was installed on the RasPi. A simple test using both a LED and a push button proved that the skill was installed successfully. A detailed review of 26 different Mycroft Skills followed the GPIO skill discussion.

The chapter concluded with an HA Skill installation using a Philips Hue light. The light was turned on and off using voice commands.

## CHAPTER 8

Fuzzy Logic and Home Automation

THIS CHAPTER WILL SHOW HOW to apply fuzzy logic (FL) concepts in an HA system. I will take a practical approach in the discussions using the RasPi in FL demonstrations while constantly introducing and explaining the different concepts and components that constitute an FL system.

A Simple HVAC FL System

This section concerns a very basic HVAC system that uses an FL controller that can only operate in either a heating or cooling mode. This is a deliberate oversimplification of a real-world system that allows me to more easily explain how FL can be integrated into a practical HA system.

Figure 8-1 is block diagram of an HVAC system that employs a traditional feedback control scheme. The entire system, as it is shown in the figure, is known as a *plant* using control system terminology. This configuration or topology is known as *feedback control* because a sensor in the living space measures the ambient room temperature, which the controller then uses to compare against a set point or target temperature. If the sensor temperature is higher than the target temperature, a cooling command is sent to the HVAC system; otherwise, a heating command is sent. In reality, there is a limited temperature region where neither heating nor cooling commands are sent. This is the *comfort zone*, where users in the living space require no interaction with the HVAC system. The FL system will handily incorporate a comfort zone.

The FL controller (FLC) replaces the traditional feedback controller in the plant. It also requires a temperature sensor, but the decision process about choosing heating,

Parts List

Item	Model	Quantity	Source
RasPi 3	B	1	adafruit.com amazon.com mcmelectronics.com
LED	Commodity	4	adafruit.com
330-ohm (Ω), $\frac{1}{8}$-watt (W) resistor	Commodity	4	adafruit.com

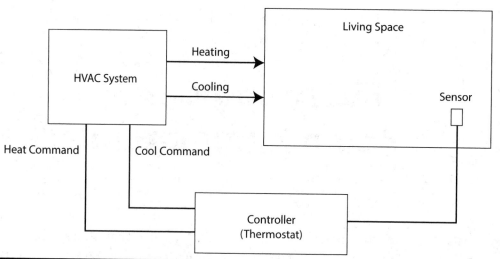

Figure 8-1   Traditional HVAC system block diagram.

cooling, or none is significantly different from a simplistic comparison with a single target value.

It will be useful at this point to first discuss some fundamental FL concepts before proceeding with an FLC demonstration.

Basic FL Concepts

FL uses the phrase *linguistic terms* to describe system or plant variables that originate in the non-FL domain. To understand what linguistic terms means, it is useful to think of how you would describe a comfortable room temperature. You might say 72°F is comfortable. However, someone else might say that 70°F is uncomfortable. In fact, it is entirely possible to survey a large, randomly selected group of people and discover that comfortable temperatures could range from 50 to 90°F. Admittedly, there would be very few people who would find the extreme temperatures to be comfortable.

Figure 8-2 shows a graph of a membership function in which the vertical axis represents membership in the range of 0 to 1.0 and temperature on the horizontal axis. A value of 1.0 at a given temperature means that it is equal to 100 percent membership within the function. Likewise, a temperature of 60°F intersects at a 0.5 membership function value, which translates to that specific temperature having a 50 percent membership function. This graph represents the translation of the plant temperature variable to the FL comfortable temperature linguistic term.

Other membership functions can be similarly created to encompass the following linguistic terms:

- Cold
- Comfortable
- Hot

It would be useful at this point to slightly change the traditional HVAC system design to one that incorporates an FLC. Figure 8-3 shows a generic FL HVAC system that incorporates

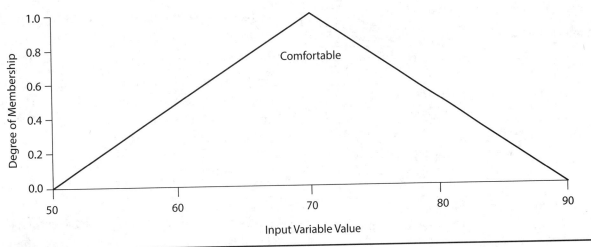

Figure 8-2 Comfortable temperature membership function.

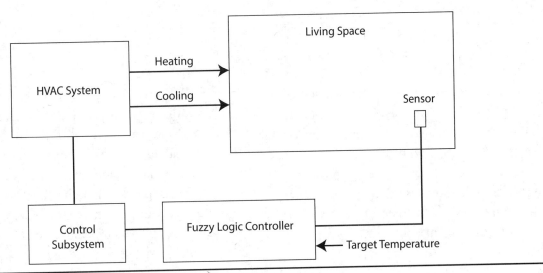

Figure 8-3 Generic FLC HVAC block diagram.

a new component—a control subsystem. The control subsystem block represents the additional hardware/software required to implement any new HVAC features resulting from using an FL approach.

There is a very specific procedure to be followed to implement an FL system, which is discussed in the next section.

FL Implementation Procedure

This implementation procedure creates an FL algorithm, which is a step-by-step process encompassing all the necessary components to generate a workable FL solution. Table 8-1 shows the seven major steps involved in creating an FL algorithm. I will explain each step in-depth after the table.

Table 8-1   Fuzzy Logic Algorithm Steps

Step	Step Name	Step Description
1	Initialization	Define FL linguistic variables and terms
2	Initialization	Generate membership functions
3	Initialization	Create expert rule set
4	Fuzzification	Convert crisp input data into fuzzy set using membership functions
5	Inference	Evaluate fuzzy set according to expert rule set
6	Aggregation	Combine rule results to form a fuzzy output set
7	Defuzzification	Convert fuzzy output set to crisp output values

STEP 1. The linguistic variables are symbols that represent system inputs and outputs. They are not numerical values but are usually natural-language words or phrases from a language such as English. Linguistic variables are further decomposed into a set of linguistic terms for both inputs and outputs. There is only one input variable, which is named `roomTemp` for this simple HVAC demonstration. There is also only one output variable, named `controlOut`, which is the command that goes to the control subsystem.

The room temperature input variable is further classified (decomposed) into a series of linguistic terms that are appropriate for the users and the HVAC environment:

- Cold
- Comfortable
- Hot

The `controlOut` output variable is further classified (decomposed) into a series of linguistic terms that are appropriate for the HVAC control subsystem:

- Heat
- No action
- Cool

STEP 2. Membership functions are prerequisites for both the fuzzification and defuzzification steps. Membership functions map nonfuzzy input values to fuzzy linguistic variables for the fuzzification step. In a similar manner, a membership function maps fuzzy variables to nonfuzzy output values for the defuzzification step. Basically, membership functions quantify linguistic terms. Figure 8-4 shows

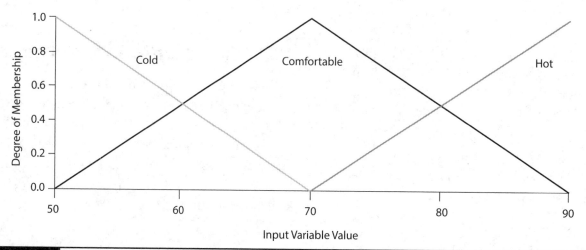

Figure 8-4   `roomTemp` membership functions.

the membership function graphs for the input variable `roomTemp`, which encompasses all the linguistic terms assigned to the input variable.

There can be many basic shapes to membership functions, but the triangular shape seems to be quite common, at least when it comes to capturing human behavior associated with the input variable.

Figure 8-5 is a graph showing the control output membership functions encompassing all three linguistic terms comprising the `controlOut` output variable. This figure has triangular membership function shapes, which is fairly common when configuring output control functions.

STEP 3. A very important component in any FL system is an expert decision system. The purpose of the expert system is to generate appropriate control actions based on the fuzzified input variables. The expert system uses the classic modus ponens form *if <condition>, then <conclusion>*, where decisions are based on the state or value of input variable(s). The following are the rules implemented for this simple HVAC demonstration:

1. If (`roomTemp` is `cold`) and (`targetTemp` is `comfortable`), the `controlOut` command is `heat`.

2. If (`roomTemp` is `hot`) and (`targetTemp` is `comfortable`), then the `controlOut` command is `cool`.

3. If (`roomTemp` is `comfortable`) and (`targetTemp` is `comfortable`), then the `controlOut` command is `no-action`.

These three rules apply both to the input and the output variables. How the rules are enforced is discussed in step 5.

STEP 4. This step requires that the crisp `roomTemp` input variable be fuzzified using the previously defined membership functions. How this happens depends on the computer language/library used in the actual FL system implementation. I will be using the Python language along with the Python SK library for this simple HVAC demonstration. I will provide detailed comments on the fuzzication process within the code listing when it is shown later in this chapter.

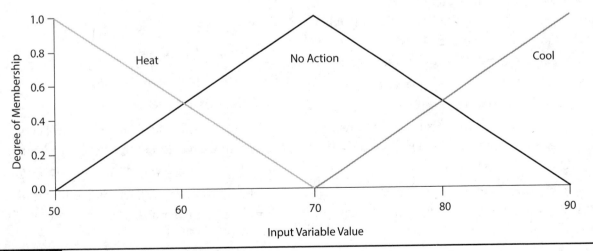

Figure 8-5   `controlOut` membership functions.

STEP 5. This step is all about applying the *if, then* inferential rules and paying attention to using only related linguistic terms. For instance, rule 1 is

If (`roomTemp` is `cold`) and (`targetTemp` is `comfortable`), the `controlOut` command is `heat`.

Applying this rule is trivial because there is only one input linguistic term that needs to be evaluated. You will see that rule application becomes more complex when additional linguistic terms are simultaneously evaluated. I will explain later the complex evaluations in the complex HVAC demonstration. For right now, this simple demonstration only relies on a one-to-one rule application. For example, if the `targetTemp` input variable is at 55°F, then, referring to Figure 8-4, you can see that the `cold` membership function value is approximately 0.7 and the `comfortable` membership function value is approximately 0.3. The highest or maximum membership value is always used when there are two membership functions being evaluated for a single crisp input variable. This means that the action associated with the `cold` membership function will be acted on using a 0.7 value for an action level to be applied to the HVAC heating mode. I will further expand on what I mean by an *action level* when I go through the complex HVAC example. For right now, simply construe the level to equate to an approximate 70 percent output variable level.

You should also easily see that an input `roomTemp` of 70°F would mean that it equates to a 1.0 `comfortable` membership, which causes the `no-action` mode to be entered for the HVAC system.

A final example of an input `roomTemp` of 80°F would mean that it equates to a 0.5

membership value for both the `comfortable` and the `hot` membership functions. In a case such as this, you would have to provide some sort of tie-breaker function to select one linguistic term over another. It probably doesn't make much of a difference regarding which term is selected, but I would likely favor the `no-action` mode versus the `cool` mode, just to economize on energy consumption.

The aggregation step is next, where all the rules have been applied to all the membership functions.

STEP 6. The `maximum` operator is normally applied to the output variable membership functions for this aggregation step. Figure 8-6 shows the combined output variable graph, which encompasses all three control modes.

There is only one more step in the FLS algorithm, and that is defuzzification.

STEP 7. *Defuzzification* is the process in which a real-world crisp output value is generated that can be acted on using the appropriate mode decided by the expert system. The following six mathematical techniques are commonly used for defuzzification:

- Centroid of area
- Bisector of area
- Smallest of maximum
- Largest of maximum
- Mean of maximum
- Weighted average

Figure 8-7 graphically demonstrates how values for each method are chosen using an arbitrary aggregation membership function.

The centroid defuzzification method is used most commonly because it is very accurate. It

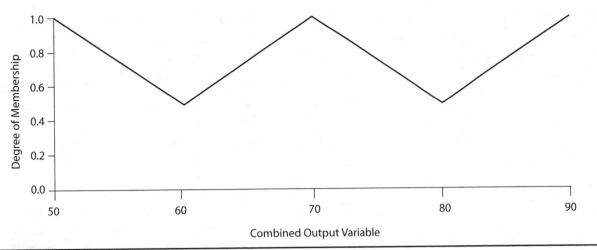

Figure 8-6 Combined output variable graph.

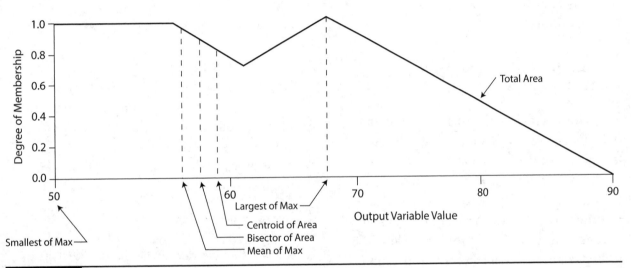

Figure 8-7 Defuzzification methods.

calculates the center of the area under the curve of the membership function. This method can require significant computational processing, especially for complex membership functions. The centroid equation is

$$z_0 = \int \mu_i(x)x\,dx / \int \mu_i(x)\,dx$$

where z_0 is the defuzzified output, μ_i represents a membership function, and x is an output variable. Bisector defuzzification uses vertical lines that divide the area under the membership curve into two equal areas where

$$\int_a^z \mu_A(x)\,dx = \int_z^\beta \mu_A(x)\,dx$$

The mean-of-maximum (MOM) defuzzification method uses the average value of the aggregated membership function outputs:

$$z_0 = \sum_{i=1}^n \frac{\omega_i}{n}$$

The smallest-of-maximum defuzzification method uses the minimum value of the aggregated membership function outputs:

$$z_0 \epsilon \{x \mid \mu(x) = \min \mu(\omega)$$

The largest-of-maximum defuzzification method uses the maximum value of the aggregated membership function outputs:

$$z_0 \epsilon \{x \mid \mu(x) = \max \mu(\omega)\}$$

The weighted-average defuzzification method calculates the weighted sum of each fuzzy set. The crisp value is set according to the weighted values and the degree of membership for fuzzy output as determined by the following formula:

$$z_0 = \frac{\Sigma \mu(x)_i W_i}{\Sigma \mu(x)_i}$$

where is the degree of membership in output singleton i, and W_i is the fuzzy output weight value for the output singleton i.

The defuzzification step completes the seven-step discussion of how to create an FL solution. The next section is a presentation of a complete Python solution for the simple HVAC demonstration.

Python Script for a Simple HVAC System

The initial step when creating a new program or script is to list the requirements or what is expected or desired to happen when the program is run (executed). Accurately specifying requirements is a key factor in how you can determine whether you have been successful in generating a workable problem and/or project solution. After careful consideration, I developed the following requirements for a Python script that would accurately simulate a simple HVAC system:

- **Two input variables.** roomTemp for the actual room temperature, and targetTemp for the desired or set-point temperature.

- **One output variable.** controlOut to send a defuzzified numeric value to a simulated control subsystem.

- **LED displays.** Three LEDs that indicate the output modes of heat, cool, and no-action.

- **Keyboard entries.** roomTemp and targetTemp.

- **Variable display.** controlOut numeric value.

You will first need to do some configurations and install some Python libraries in preparation to enter and run the script. Enter the following commands at the command line to update the Raspbian distribution and install the numpy, scipy, and matplotlib libraries:

```
sudo apt-get update
sudo apt-get install python-numpy
```

NOTE: The numpy library may already be installed, so all you will see is a message stating that the latest version is already installed.

```
sudo apt-get install python-scipy
sudo apt-get install python-matplotlib
```

The skfuzzy library, which contains all the Python fuzzy software, is somewhat more complex to install. You first need to clone the software from the GitHub website (https://github.com/scikit-fuzzy/scikit-fuzzy.git). However, you will need the git application to do this. Install git by entering this command:

```
sudo apt-get install git
```

You will then need to clone the skfuzzy software once git is installed by entering this command:

```
sudo git clone https://github.com/
    scikit-fuzzy/scikit-fuzzy.git
```

The cloning operation automatically unzips all the skfuzzy software into a new subdirectory named *scikit-fuzzy* located in the Home directory. Enter the following commands to set up the skfuzzy library:

```
cd scikit-fuzzy
sudo python setup.py install
```

You will see a lot of dialog scroll by as the skfuzzy installation progresses. You should be ready to enter and execute fuzzy Python scripts after the installation completes.

The following sections all concern going through the seven FL implementation steps. Just note that I did not follow the step order described in Table 8-1. It is not critical that the steps be done in any specific order but only that they all are eventually completed.

Generating the Expert System Rules

Six rules are required to accommodate all the combinations of room temperature and target temperature linguistic terms that require an action. All the rules for no-action are ignored. These six rules are as follows:

1. If `roomTemp` is `cold` and `targetTemp` is `comfortable`, then `controlOut` is `heat`.

2. If `roomTemp` is `cold` and `targetTemp` is `hot`, then `controlOut` is `heat`.

3. If `roomTemp` is `comfortable` and `targetTemp` is `cold`, then `controlOut` is `cool`.

4. If `roomTemp` is `comfortable` and `targetTemp` is `hot`, then `controlOut` is `heat`.

5. If `roomTemp` is `hot` and `targetTemp` is `cold`, then `controlOut` is `cool`.

6. If `roomTemp` is `hot` and `targetTemp` is `comfortable`, then `controlOut` is `cool`.

Table 8-2 is a summary matrix detailing the control commands for all combinations of linguistic variables for both room and target temperatures.

Table 8-2 Matrix of Command Actions for Room and Target Temperature Linguistic Variables

Room Temperature (`roomTemp`)	Target Temperature (`targetTemp`)		
	cold	comfortable	hot
cold	no-action	heat	heat
comfortable	cool	no-action	heat
hot	cool	cool	no-action

At this point, it is time to discuss the fuzzification step, once the rule set has been generated.

Fuzzification

The following Python code segment sets up the input variable ranges and the input and output membership functions:

```python
import numpy as np
import skfuzzy as fuzz

# Generate universe variables
# room and target temperature range is 50 to 90
# same for the output control variable
room_temp    = np.arange(50, 91, 1)
target_temp  = np.arange(50, 91, 1)
control_temp = np.arange(50, 91, 1)

# Generate triangular fuzzy membership functions
room_temp_lo    = fuzz.trimf(room_temp, [50, 50, 70])
room_temp_md    = fuzz.trimf(room_temp, [50, 70, 90])
room_temp_hi    = fuzz.trimf(room_temp, [70, 90, 90])
target_temp_lo  = fuzz.trimf(target_temp, [50, 50, 70])
```

```
target_temp_md    = fuzz.trimf(target_temp, [50, 70, 90])
target_temp_hi    = fuzz.trimf(target_temp, [50, 90, 90])
control_temp_lo = fuzz.trimf(control_temp, [50, 70, 90])
control_temp_hi = fuzz.trimf(control_temp, [70, 90, 90])
```

The next step in the algorithm is to determine the fuzzified values based on values for room and target temperatures. Based on project requirements, the user will be asked to input both values. In a real-world FL control system, the target temperature would be set manually, whereas the room temperature would be read from a sensor. However, to simply things in this simple project, both inputs will be set manually. The following code accepts user inputs and then fuzzifies those inputs:

```
# Get user inputs
room_temp_in = raw_input('Enter room temperature 50 to 90')
target_temp_in = raw_input('Enter target temperature 50 to 90')

# Calculate degrees of membership
room_temp_level_lo = fuzz.interp_membership(room_temp, room_temp_lo, float(room_temp_in))
room_temp_level_md = fuzz.interp_membership(room_temp, room_temp_md, float(room_temp_in))
room_temp_level_hi = fuzz.interp_membership(room_temp, room_temp_hi, float(room_temp_in))

target_temp_level_lo = fuzz.interp_membership(target_temp, target_temp_lo,
    float(target_temp_in))
target_temp_level_md = fuzz.interp_membership(x_target_temp, target_temp_md,
    float(target_temp_in))
target_temp_level_hi = fuzz.interp_membership(x_target_temp, target_temp_hi,
    float(target_temp_in))
```

The next step is the inference step, where all the rules are applied and membership functions combined.

Inference

The following code segment applies the six rules and combines all the membership functions:

```
# Apply rule 1:  if room_temp is cold and target temp is comfortable then command is heat
# The 'and' operator means to take the minimum by using the 'np.fmin' function
active_rule1 = np.fmin(room_temp_level_lo, target_temp_level_md)
# Combine with hi control membership function using 'np.fmin'
control_activation_1 = np.fmin(active_rule1, control_temp_hi)

# Next iterate through all five remaining rules
#Apply rule 2: if room_temp is cold and target temp is hot then command is heat
active_rule2 = np.fmin(room_temp_level_lo, target_temp_level_hi)
# Combine with hi control membership function using 'np.fmin'
control_activation_2 = np.fmin(active_rule2, control_temp_hi)
```

```
#Apply rule 3: if room_temp is comfortable and target temp is cold then command is cool
active_rule3 = np.fmin(room_temp_level_md, target_temp_level_lo)
# Combine with lo control membership function using 'np.fmin'
control_activation_3 = np.fmin(active_rule3, control_temp_lo)

#Apply rule 4: if room_temp is comfortable and target temp is heat then command is heat
active_rule4 = np.fmin(room_temp_level_md, target_temp_level_hi)
# Combine with hi control membership function using 'np.fmin'
control_activation_4 = np.fmin(active_rule4, control_temp_hi)

#Apply rule 5: if room_temp is hot and target temp is cold then command is cool
active_rule5 = np.fmin(room_temp_level_hi, target_temp_level_lo)
# Combine with lo control membership function using 'np.fmin'
control_activation_5 = np.fmin(active_rule5, control_temp_lo)


#Apply rule 6: if room_temp is hot and target temp is comfortable then command is cool
active_rule6 = np.fmin(room_temp_level_hi, target_temp_level_md)
# Combine with lo control membership function using 'np.fmin'
control_activation_6 = np.fmin(active_rule6, control_temp_lo)
```

This completes the rule application and membership set combinations. The next step to consider is the aggregation step.

Aggregation

The aggregation statement is long because of the six control activation values.

```
aggregated = np.fmax(control_activation_1,
                     control_activation_2,
                     control_activation_3,
                     control_activation_4,
                     control_activation_5,
                     control_activation_6)
```

You may notice that in the actual listing below, I had to break up this statement into a series of statements because the fmax function can only take two arguments instead of the six shown above.

It is time for the defuzzification step once the aggregation is completed.

Defuzzification

The centroid method will be used for this project. The following code defuzzifies the output control value:

```
# Calculate defuzzified result using the
#  method of centroids
control_value = fuzz.defuzz(control_temp,
     aggregated, 'centroid')
```

Now simply display the crisp output value.

```
print control_value
```

Simple HVAC System Python Script

The following is the complete listing for an initial Python script that I named *simpleHVAC.py*. It is a compilation of all the previous code snippets along with some "glue" code to ensure that all the parts function well together. However, this listing does not include any code for controlling the LEDs, which are intended to show the selected active output control mode. That script will be shown after the results of this initial script are displayed and analyzed.

```python
import numpy as np
import skfuzzy as fuzz

# Generate universe variables
#    * room and target temperature range is 50 to 90
#    * same for the output control variable
x_room_temp    = np.arange(50, 91, 1)
x_target_temp  = np.arange(50, 91, 1)
x_control_temp = np.arange(50, 91, 1)

# Generate triangular fuzzy membership functions
room_temp_lo      = fuzz.trimf(x_room_temp,    [50, 50, 70])
room_temp_md      = fuzz.trimf(x_room_temp,    [50, 70, 90])
room_temp_hi      = fuzz.trimf(x_room_temp,    [70, 90, 90])
target_temp_lo    = fuzz.trimf(x_target_temp,  [50, 50, 70])
target_temp_md    = fuzz.trimf(x_target_temp,  [50, 70, 90])
target_temp_hi    = fuzz.trimf(x_target_temp,  [50, 90, 90])
control_temp_lo   = fuzz.trimf(x_control_temp, [50, 50, 70])
control_temp_md   = fuzz.trimf(x_control_temp, [50, 70, 90])
control_temp_hi   = fuzz.trimf(x_control_temp, [70, 90, 90])

# Get user inputs
room_temp = raw_input('Enter room temperature 50 to 90: ')
target_temp = raw_input('Enter target temperature 50 to 90: ')

# Calculate degrees of membership
room_temp_level_lo = fuzz.interp_membership(x_room_temp, room_temp_lo, float(room_temp))
room_temp_level_md = fuzz.interp_membership(x_room_temp, room_temp_md, float(room_temp))
room_temp_level_hi = fuzz.interp_membership(x_room_temp, room_temp_hi, float(room_temp))

target_temp_level_lo = fuzz.interp_membership(x_target_temp, target_temp_lo,
    float(target_temp))
target_temp_level_md = fuzz.interp_membership(x_target_temp, target_temp_md,
    float(target_temp))
target_temp_level_hi = fuzz.interp_membership(x_target_temp, target_temp_hi,
    float(target_temp))

# Apply all six rules
# rule 1:  if room_temp is cold and target temp is comfortable then command is heat
active_rule1 = np.fmin(room_temp_level_lo, target_temp_level_md)
control_activation_1 = np.fmin(active_rule1, control_temp_hi)

# rule 2: if room_temp is cold and target temp is hot then command is heat
active_rule2 = np.fmin(room_temp_level_lo, target_temp_level_hi)
control_activation_2 = np.fmin(active_rule2, control_temp_hi)
```

```
# rule 3: if room_temp is comfortable and target temp is cold then command is cool
active_rule3 = np.fmin(room_temp_level_md, target_temp_level_lo)
control_activation_3 = np.fmin(active_rule3, control_temp_lo)

# rule 4: if room_temp is comfortable and target temp is heat then command is heat
active_rule4 = np.fmin(room_temp_level_md, target_temp_level_hi)
control_activation_4 = np.fmin(active_rule4, control_temp_hi)

# rule 5: if room_temp is hot and target temp is cold then command is cool
active_rule5 = np.fmin(room_temp_level_hi, target_temp_level_lo)
control_activation_5 = np.fmin(active_rule5, control_temp_lo)

# rule 6: if room_temp is hot and target temp is comfortable then command is cool
active_rule6 = np.fmin(room_temp_level_hi, target_temp_level_md)
control_activation_6 = np.fmin(active_rule6, control_temp_lo)

# Aggregate all six output membership functions together
# Combine outputs to ease the complexity as fmax() only as two args
c1 = np.fmax(control_activation_1, control_activation_2)
c2 = np.fmax(control_activation_3, control_activation_4)
c3 = np.fmax(control_activation_5, control_activation_6)
c4 = np.fmax(c2,c3)
aggregated = np.fmax(c1, c4)

# Calculate defuzzified result using the method of centroids
control_value = fuzz.defuzz(x_control_temp, aggregated, 'centroid')

#  Display the crisp output value
print control_value
```

Testing the Simple HVAC System Script

Run the script by entering the following command at the terminal prompt in the home directory:

```
python simpleHVAC.py
```

Table 8-3 shows the results of testing the control script using a representative range of room and target temperatures that were input manually. Note that all table entries are in °F.

I carefully studied these results and derived these conclusions from the test data:

■ A command value of approximately 65 to 75 means no action.

■ A command value of approximately 82 to 83 means that heating is required.

■ A command value of approximately 56 to 65 means that cooling is required.

The no action range was approximately ±4 surrounding the target temperature. This is a good result because it minimizes system operation while still maintaining the desired room temperature.

Table 8-3   Simple HVAC System Script Test Results

Target Temperature	Room Temperature	Command Output
50	50	70.00
	60	57.78
	70	56.67
	80	57.78
	90	56.67
60	50	82.22
	60	70.00
	70	66.40
	80	66.40
	90	57.78
70	50	83.33
	60	82.22
	70	82.22
	80	70.00
	90	56.67
80	50	83.33
	60	82.22
	70	83.33
	80	70.00
	90	57.78
90	50	83.33
	60	82.22
	70	83.33
	80	82.22
	90	70.00

LED Mode Indicators

I next made a few simple modifications to the simple HVAC system script that would light one of three LEDs depending on whether heating, cooling, or no action were determined based on user input. You should use the Fritzing diagram in Figure 8-8 to set up the LEDs to indicate the selected control mode.

The following listing incorporates the LED modifications to the existing Simple HVAC system script (*simpleHVAC.py*). I renamed it *simpleHVAC_LED.py*, and it is available from this book's companion website, www.mhprofessional .com/NorrisHomeAutomation.

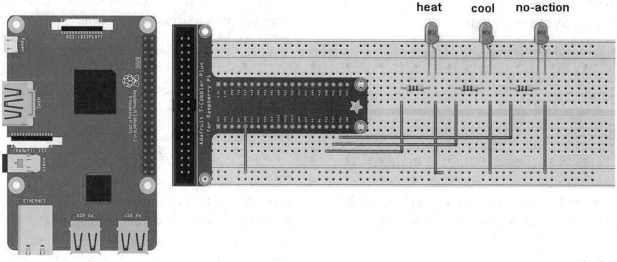

Figure 8-8   Fritzing diagram for control mode indicator LEDs.

```python
import numpy as np
import skfuzzy as fuzz
import RPi.GPIO as GPIO
import time

GPIO.setmode(GPIO.BCM)

# prevent unnecessary GPIO warnings when re-running the script
GPIO.setwarnings(False)

# configure BCM GPIO pins 13, 19 and 26 as outputs
GPIO.setup(13, GPIO.OUT)
GPIO.setup(19, GPIO.OUT)
GPIO.setup(26, GPIO.OUT)

# set all LEDs to LOW
GPIO.output(13, GPIO.LOW)
GPIO.output(19, GPIO.LOW)
GPIO.output(26, GPIO.LOW)

# Generate universe variables
# room and target temperature range is 50 to 90
# same for the output control variable
x_room_temp   = np.arange(50, 91, 1)
x_target_temp = np.arange(50, 91, 1)
x_control_temp = np.arange(50, 91, 1)

# Generate fuzzy triangular membership functions
room_temp_lo    = fuzz.trimf(x_room_temp,    [50, 50, 70])
room_temp_md    = fuzz.trimf(x_room_temp,    [50, 70, 90])
room_temp_hi    = fuzz.trimf(x_room_temp,    [70, 90, 90])
target_temp_lo  = fuzz.trimf(x_target_temp, [50, 50, 70])
target_temp_md  = fuzz.trimf(x_target_temp, [50, 70, 90])
target_temp_hi  = fuzz.trimf(x_target_temp, [50, 90, 90])
control_temp_lo = fuzz.trimf(x_control_temp,[50, 50, 70])
control_temp_md = fuzz.trimf(x_control_temp,[50, 70, 90])
control_temp_hi = fuzz.trimf(x_control_temp,[70, 90, 90])


# Get user inputs
room_temp = raw_input('Enter room temperature 50 to 90: ')
target_temp = raw_input('Enter target temperature 50 to 90: ')

# Calculate degrees of membership
room_temp_level_lo = fuzz.interp_membership(x_room_temp, room_temp_lo, float(room_temp))
room_temp_level_md = fuzz.interp_membership(x_room_temp, room_temp_md, float(room_temp))
room_temp_level_hi = fuzz.interp_membership(x_room_temp, room_temp_hi, float(room_temp))
```

```
target_temp_level_lo = fuzz.interp_membership(x_target_temp, target_temp_lo,
    float(target_temp))
target_temp_level_md = fuzz.interp_membership(x_target_temp, target_temp_md,
    float(target_temp))
target_temp_level_hi = fuzz.interp_membership(x_target_temp, target_temp_hi,
    float(target_temp))

# Apply all six rules
# rule 1:  if room_temp is cold and target temp is comfortable then command is heat
active_rule1 = np.fmin(room_temp_level_lo, target_temp_level_md)
control_activation_1 = np.fmin(active_rule1, control_temp_hi)

# rule 2: if room_temp is cold and target temp is hot then command is heat
active_rule2 = np.fmin(room_temp_level_lo, target_temp_level_hi)
control_activation_2 = np.fmin(active_rule2, control_temp_hi)

# rule 3: if room_temp is comfortable and target temp is cold then command is cool
active_rule3 = np.fmin(room_temp_level_md, target_temp_level_lo)
control_activation_3 = np.fmin(active_rule3, control_temp_lo)

# rule 4: if room_temp is comfortable and target temp is heat then command is heat
active_rule4 = np.fmin(room_temp_level_md, target_temp_level_hi)
control_activation_4 = np.fmin(active_rule4, control_temp_hi)

# rule 5: if room_temp is hot and target temp is cold then command is cool
active_rule5 = np.fmin(room_temp_level_hi, target_temp_level_lo)
control_activation_5 = np.fmin(active_rule5, control_temp_lo)

# rule 6: if room_temp is hot and target temp is comfortable then command is cool
active_rule6 = np.fmin(room_temp_level_hi, target_temp_level_md)
control_activation_6 = np.fmin(active_rule6, control_temp_lo)

# Aggregate all six output membership functions together
# Combine outputs to ease the complexity as fmax() only as two args
c1 = np.fmax(control_activation_1, control_activation_2)
c2 = np.fmax(control_activation_3, control_activation_4)
c3 = np.fmax(control_activation_5, control_activation_6)
c4 = np.fmax(c2,c3)
aggregated = np.fmax(c1, c4)

# Calculate defuzzified result using the method of centroids
control_value = fuzz.defuzz(x_control_temp, aggregated, 'centroid')

#  Display the crisp output value
print control_value

# following limit values taken from test data analysis
# no-action mode
if control_value > 68 and control_value < 82:
    GPIO.output(26, GPIO.HIGH)
```

```
    time.sleep(5)
    GPIO.output(26, GPIO.LOW)

# heat mode
elif control_value > 82 and control_value < 84:
    GPIO.output(13, GPIO.HIGH)
    time.sleep(5)
    GPIO.output(13, GPIO.LOW)

# cool mode
elif control_value > 56 and control_value < 68:
    GPIO.output(19, GPIO.HIGH)
    time.sleep(5)
    GPIO.output(19, GPIO.LOW)

# error, default to the no-action mode
else:
    print 'out of range value calculated'
    print 'no-action mode'
    GPIO.output(26, GPIO.HIGH)
    time.sleep(5)
    GPIO.output(26, GPIO.LOW)
```

Figure 8-9 shows the physical setup with the three control LEDs connected to a solderless breadboard using a T-Cobbler interface adapter.

Test Run for the Mode Indicator LEDs

I used a set of input values taken from Table 8-3 to test the LED mode indication system. The values used and the appropriate associated modes are shown in Table 8-4.

Figure 8-9   Physical setup.

Table 8-4 LED Indicator Test Values

Mode	Room Temperature	Target Temperature	Control Value
Heat	60	80	82.22
Cool	80	60	66.40
No action	60	60	70.00

The modified script is run by entering the following command at the terminal prompt in the home directory:

```
python simpleHVAC_LED.py
```

I ran the script three times, entering the values for each control mode, and subsequently confirmed that the LED for that mode lighted for 5 seconds as expected.

This test completes the simple HVAC system demonstration. However, it may be apparent that the GPIO control design can be adapted to operate real-world HVAC control lines. The easiest way is to just substitute an optocoupler for a LED. Figure 8-10 shows an example schematic for this change.

In this circuit, a RasPi GPIO 3.3-V line is connected directly to an optocoupler, which has it output connected to a single-pole, single-throw (SPST) relay. The relay contacts control a 24-VAC line, which is the typical control voltage used in HVAC systems. You could even control

120/240-VAC HVAC lines, if necessary, using this scheme, provided that the relay contacts are rated to handle the higher voltage.

The simple HVAC system demonstration has now been completed, and you should have gained some insight into and appreciation for how FL can effectively implement a good solution to a common application such as HVAC. The next demonstration expands this simple project by introducing additional complexity to more closely model real-world HVAC applications.

Complex HVAC System Demonstration

The preceding simple demonstration used only two input parameters to control an HVAC system that had three operational modes. This situation is really an oversimplification of what is possible to achieve using FL technology. There are a number of other factors, both inputs and outputs, that have an impact on the operations of a real-world HVAC system including, but not limited to

- Humidity
- Time of day
- Occupancy

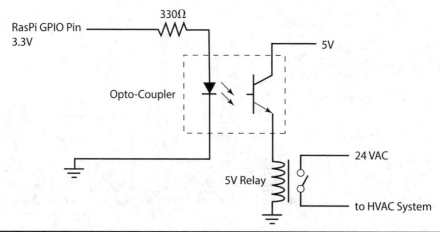

Figure 8-10 Optocoupler control schematic.

- Supply airflow
- Exhaust airflow
- Makeup air
- Outside air temperature and humidity
- Economizer modes

You likely realize that these factors interact with each other in complex ways and that it is not a trivial task to generate appropriate membership function and expert system rules. In fact, the degree of complexity goes up exponentially with every new factor introduced into the FL control scheme. In the simple system demonstration, six rules were required to accommodate all combinations of the input and output linguistic terms that required an action. There theoretically should have been nine rules, but I chose to ignore three rules, which were dedicated to the no action mode. Now, if one more linguistic term had been added, let's say one dealing with humidity, that would have created the need for 25 rules. This large number is due to the added humidity input and a separate control output—hence $5^2 = 25$. It is not hard to imagine the combinatorial "explosion" resulting from adding many of the factors just listed to an FL system. Generating all the associated rules for many factors is a nontrivial task, which is why it is called an *expert rule set*. Experts are really the only people having all the knowledge on how the many system factors interact and what control actions are required to achieve the desired results. Even experts will have a difficult time in detailing hundreds of rules without errors or conflicts, which is why complex HVAC system design is a challenging endeavor.

Although I am not an HVAC expert, I did attempt to modify the simple HVAC system demonstration by adding a humidity input variable. Creating the linguistic terms and membership functions was relatively simple because many data are available on the Internet. Creating the expert rules was another issue,

but I did find some online data regarding how to control humidity in an HVAC system. I finally decided that I would simply send control signals to a dehumidification diverter valve to control room humidity levels. I explain in the next section how the valve works, along with some other important background information on how humidity affects people. Applying this information is key to creating meaningful membership functions as well as functional and useful expert rules.

Humidity Control

Humidity levels have a significant effect on people occupying an air-conditioned space. Humidity levels are quantified using the *dew point*, which is commonly defined as "the temperature to which air must be cooled to become saturated with water vapor." If air at the dew point is further cooled, the airborne water vapor will condense to form liquid water (dew). When air cools to its dew point through contact with a surface that is colder than the air, water will condense on the surface.

Relative humidity is often confused with dew point. *Relative humidity* (RH) is defined as "the ratio between the current amount of water vapor in the air at a given temperature and the maximum amount of water vapor possible in the air at that temperature." It has units of percent, whereas dew point has temperature units. High humidity in a building can have severe adverse effects beyond personal discomfort. Condensation can occur in building cavities that would promote mold growth and/ or other moisture damage to the building or its furnishings.

The dew point is always lower than or equal to the ambient air temperature, which is why dew or fog often occurs during the early morning hours when the air temperatures are typically lowest and the dew point highest. Dew point is a very useful measure for personal comfort and has been quantified as shown in Table 8-5. It should clear from the table values that a dew point level 55 or less should always be proper objective in an HVAC system.

Table 8-5   Dew Point versus Personal Comfort

Dew Point (°F)	Human Comfort
<50	Pleasant
50–55	Comfortable
56–60	Noticeable
61–65	Sticky
66–70	Uncomfortable
71–75	Harsh
76+	Severe discomfort

Augmented Simple HVAC System

This HVAC software demonstration assumes that a dehumidification subsystem has been installed in the plant and may be independently controlled aside from the heating and cooling subsystems. This would mean that even if the room temperature was in the comfortable

membership function, the dehumidification could be activated if the room dew point was in an uncomfortable range. Figure 8-11 shows a representative generic air-conditioning system diagram that has a dehumidification feature.

Notice in the figure that there is a dehumidification diverter valve that can divert the cool airflow coming from the inside heat exchanger to the hot airflow coming from the outside condenser. While this seems illogical at first glance, it does make sense because warming the cool air slightly will reduce the overall moisture content, thus lowering the dew point. Many modern air-conditioning systems provide this feature and call it a "dry" mode of operation. In reality, it is a dehumidification operation.

The augmented system would need an additional sensor to measure dew point levels within the room. Then, if the dew point was at an undesirable level, all appropriate rules would be invoked. These rules would likely include standalone dehumidification, which would mean just operating a diverter valve to heat

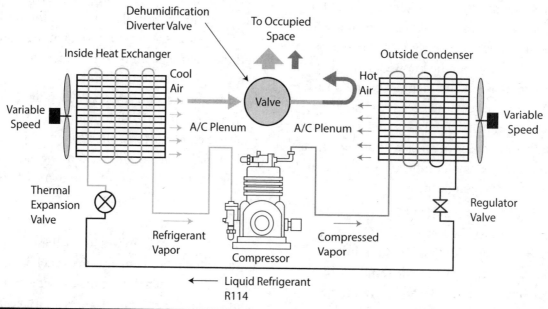

Figure 8-11   Air-conditioning system.

the cooled air in the air-conditioner plenum so that the dew point could naturally rise due to the increased capacity of the warmer air to retain moisture without it condensing. I added a few rules dealing with dehumidification just to illustrate the process, but I am no HVAC expert, as previously stated, and will have most likely left out some important rules. In any case, the software will query the user to enter both the room and target dew points, in addition to both room and target temperatures. I now need to expand the FL membership functions and add several expert rules before demonstrating a complex HVAC system simulation.

Additional Membership Functions and Expert Rules

Figure 8-12 shows a typical humidity membership function graph, which reflects how people react to humid conditions. The input variable for this graph is the dew point, and the membership values shown are in accord with the values shown in Table 8-5. Three triangular membership functions are shown in the figure, which are quite similar to the three original room temperature membership functions.

I next needed to create some rules on how to handle the new dew point input variable. After some considerable thought, I decided that a straightforward solution was best to control the room humidity. In case of too high humidity, the only output control available would be to open the dehumidification diverter valve and direct already cooled air to the warm output condenser airflow, as I said in the preceding section. The only caution I had was that the room should already been cooled or heated to the target temperature. Otherwise, the system could be set in an unstable state where neither the room temperature nor the desired dew point could be achieved. This prerequisite meant that the target temperature was already reached and that the no action mode was in effect. Therefore, the dehumidification would only be enabled provided that the system was in the no action mode.

I created two dehumidification rules based on the preceding discussion. These are

1. If the dew point is *high* and the target dew point is *comfortable*, then the command is `dehumidify`.

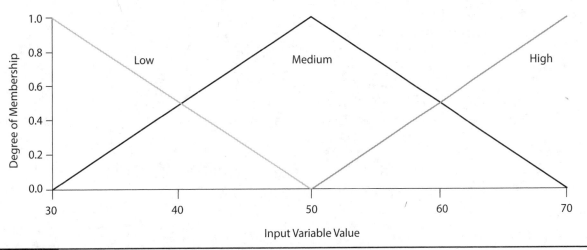

Figure 8-12 Humidity membership functions.

2. If the dew point is *comfortable* and target dew point is *low*, then the command is `dehumidify`.

I also set up another LED to indicate when the dehumidification mode started.

Complex HVAC System Python Script

The simpleHVAC_LED script was modified extensively, incorporating the new rules and membership functions just discussed. All the existing temperature control features are still in place and functioning. The new script is named *complexHVAC.py* and is available from this book's website, www.mhprofessional.com/NorrisHomeAutomation.

```python
import numpy as np
import skfuzzy as fuzz
import RPi.GPIO as GPIO
import time

GPIO.setmode(GPIO.BCM)

GPIO.setwarnings(False)

GPIO.setup(13, GPIO.OUT)
GPIO.setup(19, GPIO.OUT)
GPIO.setup(26, GPIO.OUT)
GPIO.setup(21, GPIO.OUT)

GPIO.output(13, GPIO.LOW)
GPIO.output(19, GPIO.LOW)
GPIO.output(26, GPIO.LOW)
GPIO.output(21, GPIO.LOW)

# Generate universe variables
# room and target temperature range is 50 to 90
# same for the output control variable
x_room_temp    = np.arange(50, 91, 1)
x_target_temp  = np.arange(50, 91, 1)
x_control_temp = np.arange(50, 91, 1)

# Generate humidity variables
x_dew_point        = np.arange(30, 71, 1)
x_target_dew_point = np.arange(30, 71, 1)
x_control_humid    = np.arange(30, 71, 1)

# Generate fuzzy membership functions
room_temp_lo    = fuzz.trimf(x_room_temp,    [50, 50, 70])
room_temp_md    = fuzz.trimf(x_room_temp,    [50, 70, 90])
room_temp_hi    = fuzz.trimf(x_room_temp,    [70, 90, 90])

target_temp_lo  = fuzz.trimf(x_target_temp, [50, 50, 70])
target_temp_md  = fuzz.trimf(x_target_temp, [50, 70, 90])
target_temp_hi  = fuzz.trimf(x_target_temp, [50, 90, 90])
```

```
control_temp_lo  = fuzz.trimf(x_control_temp,[50, 50, 70])
control_temp_md  = fuzz.trimf(x_control_temp,[50, 70, 90])
control_temp_hi  = fuzz.trimf(x_control_temp,[70, 90, 90])

# Generate humidity control membership functions
dew_point_lo       = fuzz.trimf(x_dew_point,  [30, 30, 50])
dew_point_md       = fuzz.trimf(x_dew_point,  [30, 50, 70])
dew_point_hi       = fuzz.trimf(x_dew_point,  [50, 70, 70])

target_dew_point_lo = fuzz.trimf(x_target_dew_point, [30, 30, 50])
target_dew_point_md = fuzz.trimf(x_target_dew_point, [30, 50, 70])
target_dew_point_hi = fuzz.trimf(x_target_dew_point, [50, 70, 70])

control_dew_point_lo = fuzz.trimf(x_control_humid, [30, 30, 50])
control_dew_point_md = fuzz.trimf(x_control_humid, [30, 50, 70])
control_dew_point_hi = fuzz.trimf(x_control_humid, [50, 70, 70])

# Get user inputs
room_temp = raw_input('Enter room temperature 50 to 90: ')
target_temp = raw_input('Enter target temperature 50 to 90: ')
dew_point = raw_input('Enter room dew point 20 to 80: ')
target_dew_point = raw_input('Enter target dew point 20 to 80: ')

# Calculate degrees of membership
room_temp_level_lo = fuzz.interp_membership(x_room_temp, room_temp_lo, float(room_temp))
room_temp_level_md = fuzz.interp_membership(x_room_temp, room_temp_md, float(room_temp))
room_temp_level_hi = fuzz.interp_membership(x_room_temp, room_temp_hi, float(room_temp))

target_temp_level_lo = fuzz.interp_membership(x_target_temp, target_temp_lo,
float(target_temp))
target_temp_level_md = fuzz.interp_membership(x_target_temp, target_temp_md,
float(target_temp))
target_temp_level_hi = fuzz.interp_membership(x_target_temp, target_temp_hi,
float(target_temp))

dew_point_level_lo = fuzz.interp_membership(x_dew_point, dew_point_lo, float(dew_point))
dew_point_level_md = fuzz.interp_membership(x_dew_point, dew_point_md, float(dew_point))
dew_point_level_hi = fuzz.interp_membership(x_dew_point, dew_point_hi, float(dew_point))

target_dew_point_level_lo = fuzz.interp_membership(x_target_dew_point,
    target_dew_point_lo, float(target_dew_point))
target_dew_point_level_md = fuzz.interp_membership(x_target_dew_point,
    target_dew_point_md, float(target_dew_point))
target_dew_point_level_hi = fuzz.interp_membership(x_target_dew_point,
    target_dew_point_hi, float(target_dew_point))

# Apply all six rules for temprature control
# rule 1:  if room_temp is cold and target temp is comfortable then command is heat
active_rule1 = np.fmin(room_temp_level_lo, target_temp_level_md)
control_activation_1 = np.fmin(active_rule1, control_temp_hi)
```

```
# rule 2: if room_temp is cold and target temp is hot then command is heat
active_rule2 = np.fmin(room_temp_level_lo, target_temp_level_hi)
control_activation_2 = np.fmin(active_rule2, control_temp_hi)

# rule 3: if room_temp is comfortable and target temp is cold then command is cool
active_rule3 = np.fmin(room_temp_level_md, target_temp_level_lo)
control_activation_3 = np.fmin(active_rule3, control_temp_lo)

# rule 4: if room_temp is comfortable and target temp is heat then command is heat
active_rule4 = np.fmin(room_temp_level_md, target_temp_level_hi)
control_activation_4 = np.fmin(active_rule4, control_temp_hi)

# rule 5: if room_temp is hot and target temp is cold then command is cool
active_rule5 = np.fmin(room_temp_level_hi, target_temp_level_lo)
control_activation_5 = np.fmin(active_rule5, control_temp_lo)

# rule 6: if room_temp is hot and target temp is comfortable then command is cool
active_rule6 = np.fmin(room_temp_level_hi, target_temp_level_md)
control_activation_6 = np.fmin(active_rule6, control_temp_lo)

# Aggregate all six output membership functions together
# Combine outputs to ease the complexity as fmax() only as two args
c1 = np.fmax(control_activation_1, control_activation_2)
c2 = np.fmax(control_activation_3, control_activation_4)
c3 = np.fmax(control_activation_5, control_activation_6)
c4 = np.fmax(c2,c3)
aggregated = np.fmax(c1, c4)

# Calculate defuzzified result using the method of centroids
control_value = fuzz.defuzz(x_control_temp, aggregated, 'centroid')

#  Display the crisp output value
print 'Temp control value = ', control_value

# dehumidification rules
# rule 7: if dew point is high and target dew point is comfortable then command is dehumidify
active_rule7 = np.fmin(dew_point_level_hi, target_dew_point_md)
control_activation_7 = np.fmin(active_rule7, control_dew_point_lo)

#rule 8: if dew point is comfortable and target dew point is low then command is dehumidify
active_rule8 = np.fmin(dew_point_level_md, target_dew_point_lo)
control_activation_8 = np.fmin(active_rule8, control_dew_point_lo)

# Aggregate  the two dew point output functions
aggregate_dp = np.fmax(control_activation_7, control_activation_8)

# Defuzzify the dew point control
control_dp = fuzz.defuzz(x_control_humid, aggregate_dp, 'centroid')

# Display the dew point control value
```

```
print 'Dew point control value = ', control_dp

# no-action mode
if control_value > 68 and control_value < 82:
    GPIO.output(26, GPIO.HIGH)
    time.sleep(5)
    GPIO.output(26, GPIO.LOW)

# cool mode
elif control_value > 82 and control_value < 84:
    GPIO.output(13, GPIO.HIGH)
    time.sleep(5)
    GPIO.output(13, GPIO.LOW)

# heat mode
elif control_value > 56 and control_value < 68:
    GPIO.output(19, GPIO.HIGH)
    time.sleep(5)
    GPIO.output(19, GPIO.LOW)

# error, default to no-action mode
else:
    print 'out of range value calculated'
    print 'no-action mode'
    GPIO.output(26, GPIO.HIGH)
    time.sleep(5)
    GPIO.output(26, GPIO.LOW)

# dehumidification indicator
if control_dp > 37 and control_value > 68 and control_value < 82:
    GPIO.output(21, GPIO.HIGH)
    time.sleep(5)
    GPIO.output(21, GPIO.LOW)
```

Test Run for the Complex HVAC System Script

I used a set of input values taken from Table 8-4 to test the Complex HVAC system. The values used and the appropriate associated modes are shown in Table 8-6.

The modified script is run by entering the following command at the terminal prompt in the Home directory:

```
python complexHVAC.py
```

Table 8-6 Complex HVAC System Script Test Values

Mode	Room Temp	Target Temp	Control Value	Room Dew Point	Target Dew Point	Dew Point Control Value
Heat	60	80	82.22	60	50	n/a
Cool	80	60	66.40	60	50	n/a
No action	60	60	70.00	60	50	37.67

I ran the script three times, confirming that each mode functioned correctly, as it did in the simpleHVAC_LED script. In addition, I entered both room and target dew points for each mode. The system did not light the dehumidification LED for either the heat or cool modes, as it was designed to do. It did light when the no action mode was active. That mode's LED also lighted for 5 seconds as expected.

This test completed the complex HVAC system demonstration.

Summary

This was a chapter on how to apply fuzzy logic (FL) concepts to HVAC applications. I divided the chapter into two major parts, one dealing with a simple, temperature-controlled HVAC system and the other one describing a more complex system.

I began the first part by reviewing basic FL concepts, which are key to understanding how FL works. There was a comprehensive discussion regarding the FL implementation procedure, which I closely followed to create an FL solution on a RasPi.

A complete Python script for a simple HVAC system that runs on a RasPi was presented next. I went through all the necessary configuration and installation steps required to get the RasPi ready to run the script. The script was then successfully tested.

I next modified the script to enable a set of LEDs to indicate which HVAC mode was operational based on user inputs. This modification was also successfully tested.

The second part of the chapter dealt with designing and running a complex FL HVAC system. This was basically the same as the simple system except that many more input and output variables were necessary. I chose to add only one input and one output variable to the existing system because they were sufficient to demonstrate the issues with implementing an HVAC system with multiple I/O variables.

A modified Python script was written that incorporates dehumidification input and output variables. I explained how dehumidification can be implemented in a real HVAC system. The script simulated this action by lighting a LED when the dehumidification mode was activated.

## CHAPTER 9

Sensors

I CONSIDERED TITLING this chapter "Nuts and Bolts" because it is all about a variety of HA sensors that I believe make up the nuts and bolts of an HA system. I have discussed various sensors in previous chapters but really didn't provide too much discussion on how they work or are interfaced with a RasPi. This chapter provides the necessary background on how to connect and program several commonly used sensors found in HA systems. Hopefully, it will also provide enough guidance for you to be able to connect different sensors using similar interface connections.

Temperature and Humidity Sensors

I will cover a variety of temperature and humidity sensors to provide you with a good understanding of how these sensors function. Each sensor provides both temperature and humidity readings, but they provide the data to a RasPi using different data protocols. You will gain a good understanding of the different interface protocols by reading all these sensor discussions. You should also consider duplicating one or more of the demonstrations to further improve your knowledge and confidence in using this sensor type.

Parts List

Item	Model	Quantity	Source
RasPi 3	B	1	adafruit.com amazon.com mcmelectronics.com
Temp/humidity sensor	DHT11	1	adafruit.com
10-kiloohm (kΩ) resistor	Commodity	1	adafruit.com
Temperature sensor	TMP36	1	adafruit.com
PIR sensor	555-28027	1	parallax.com
Ultrasonic sensor	HC-SR04	1	amazon.com
Level shifter	1875	1	adafruit.com

DHT11

A number of fairly inexpensive temperature and humidity sensors are readily available for purchase. They fall into one of three categories based on how they are designed:

- **Capacitive.** These are thin-film capacitance-based sensors with an element bonded to a monolithic circuit that provides a voltage output as a function of relative humidity.

- **Resistive.** These measure the change in electrical impedance of a hygroscopic medium such as a conductive polymer, salt, or treated substrate.

- **Thermal conductivity.** These consist of two matched negative temperature coefficient

(NTC) thermistor elements in a bridge circuit; one is hermetically encapsulated in dry nitrogen, and the other is exposed to the environment.

I will be using the DHT11 resistive sensor because it is commonly used, quite inexpensive, and has good Python libraries readily available. Figure 9-1 shows two DHT11 versions that can be easily purchased. The unit on the left is the board version, and the unit on the right is the standalone component version. I used the component version in this chapter's demonstration.

The DHT11 specifications are

- Very low cost
- 3 to 5 V of power

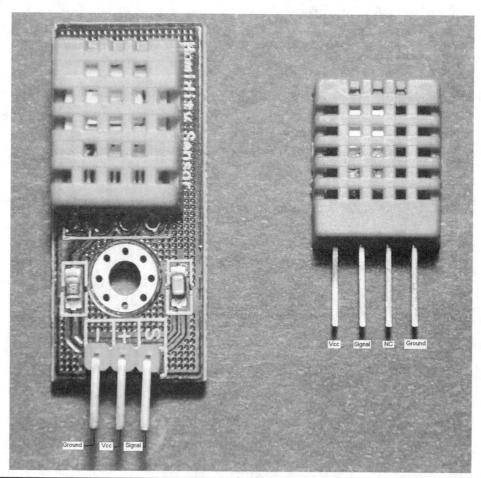

Figure 9-1 DHT11 versions.

- 3- to 5-V I/O levels

- 2.5 mA of maximum current use during conversion or requesting data

- 20 to 80 percent humidity readings with ±5 percent accuracy

- 0 to 50°C temperature readings with ±2°C accuracy

- Maximum of 1-Hz sampling rate

- Size 15.5 × 12 × 5.5 mm (component version)

- Four I/O pins with 0.1-inch spacing

There is only one signal lead on the sensor, which functions as both an input and output for digital signals. This arrangement is often referred to as a *one-wire protocol* because it only uses one signal lead for both input and output data transmission. Unfortunately, this is often confused with the 1-Wire Protocol, which is another single-wire I/O data transmission protocol. The 1-Wire Protocol is a device communications bus system designed by the Dallas Semiconductor Corp. that provides low-speed data, signaling, and power over a single wire. Both single-wire protocols perform I/O operations, but they use different timing and data-level representations. The software library for the one-wire protocol has been specifically designed to operate with only that protocol.

Physical Setup

Figure 9-2 is a Fritzing diagram showing how to connect a DHT11 with a RasPi. It is very important to connect a 10-kΩ pull-up resistor between the signal lead and the 5-V supply. The one-wire protocol will not work without this resistor installed.

Note that the supply to the sensor is 5 V, but the signal output level is 3.3 V, making the sensor completely compatible with the RasPi's GPIO pin voltage levels. The DHT11 signal-out lead connects to RasPi GPIO pin 4.

Software Installation

You should follow this terminal command sequence to install the software required to operate a DHT11 sensor using Python:

1. Installs the git application, which is required for the cloning operation:

```
sudo apt-get install git-core
```

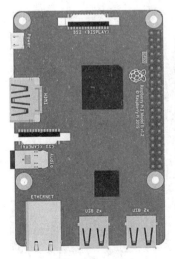
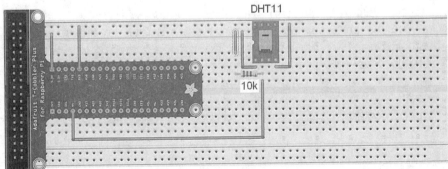

Figure 9-2   Fritzing diagram for connecting a DHT11 with a RasPi.

2. Clone the Adafruit DHT11 software from the GitHub website:

```
git clone https://githb.com/adafruit/
    Adafruit_Python_DHT.git
```

3. Change into a new directory created from the clone operation:

```
cd Adafruit_Python_DHT
```

4. Build the software:

```
sudo apt-get install build-essential
    python-dev
```

5. Run the setup script:

```
sudo python setup.py install
```

There is a file named *AdafruitDHT.py* in the Examples directory, which is located in the Adafruit_Python_DHT directory. This file will be used as a script to demonstrate the sensor's operation. You should first modify this file using the nano editor to incorporate the changes I have noted in the following listing. The modifications cause the script to continuously run as well as display temperatures in °F. You can leave out that last change if you prefer to display temperatures in °C.

```
#!/usr/bin/python
# Copyright (c) 2014 Adafruit Industries
# Author: Tony DiCola
# Modified by D. J. Norris, 2018
# Permission is hereby granted, free of charge, to any person
# obtaining a copy of this software and associated documentation files
# ( the "Software" ), to deal in the Software without restriction,
# including without limitation the rights to use, copy, modify, merge,
# publish, distribute, sublicense, and/or sell copies of the Software,
# and to permit persons to whom the Software is furnished to do so,
# subject to the following conditions:

# The above copyright notice and this permission notice shall be
# included in all copies or substantial portions of the Software.

# THE SOFTWARE IS PROVIDED "AS IS", WITHOUT WARRANTY OF ANY KIND,
# EXPRESS OR IMPLIED, INCLUDING BUT NOT LIMITED TO THE WARRANTIES OF
# MERCHANTABILITY, FITNESS FOR A PARTICULAR PURPOSE AND
# NONINFRINGEMENT. IN NO EVENT SHALL THE AUTHORS OR COPYRIGHT HOLDERS
# BE LIABLE FOR ANY CLAIM, DAMAGES OR OTHER LIABILITY, WHETHER IN AN
# ACTION OF CONTRACT, TORT OR OTHERWISE, ARISING FROM, OUT OF OR IN
# CONNECTION WITH THE SOFTWARE OR THE USE OR OTHER DEALINGS IN THE
# SOFTWARE.

import sys
import Adafruit_DHT
# added - DJN
import time
```

```
# Parse command line parameters.
sensor_args = { '11': Adafruit_DHT.DHT11,
                '22': Adafruit_DHT.DHT22,
                '2302': Adafruit_DHT.AM2302 }
if len(sys.argv) == 3 and sys.argv[1] in sensor_args:
    sensor = sensor_args[sys.argv[1]]
    pin = sys.argv[2]
else:
    print('Usage: sudo ./Adafruit_DHT.py [11|22|2302] <GPIO pin number>')
    print('Example: sudo ./Adafruit_DHT.py 2302 4 - Read from an AM2302 connected to
      GPIO pin #4')
    sys.exit(1)

# Try to grab a sensor reading.  Use the read_retry method which will retry up
# to 15 times to get a sensor reading (waiting 2 seconds between each retry).
humidity, temperature = Adafruit_DHT.read_retry(sensor, pin)

# Un-comment the line below to convert the temperature to Fahrenheit.
# Uncommented - DJN
temperature = temperature * 9/5.0 + 32

# Note that sometimes you won't get a reading and
# the results will be null (because Linux can't
# guarantee the timing of calls to read the sensor).
# If this happens try again!
# added - DJN,  also pay attention to new indentations
while True:
    if humidity is not None and temperature is not None:
        print('Temp={0:0.1f}  Humidity={1:0.1f}%'.format(temperature, humidity))
# added - DJN, you change time delay to any value you desire
        time.sleep(120)
    else:
        print('Failed to get reading. Try again!')
        sys.exit(1)
```

Test Run

I ran the test script by entering the following terminal commands:

```
cd Adafruit_Python_DHT/examples
python AdafruitDHT.py 11 4
```

Figure 9-3 shows the script output after running for several hours.

TMP36

Analog Devices' TMP36 is my favorite temperature sensor, mainly because it is quite accurate and also extremely inexpensive. It is shown in Figure 9-4.

The TMP36 is housed in a standard TO-92 form factor, which is also common to most plastic-encased transistors. The TMP36 is

pi@raspberrypi: ~/Adafruit_Python_DHT/examples

File   Edit   Tabs   Help

```
Temp = 75.2  Humidity = 13.0%
Temp = 75.2  Humidity = 13.0%
Temp = 75.2  Humidity = 13.0%
Temp = 75.2  Humidity = 13.0%
Temp = 75.2  Humidity = 13.0%
Temp = 75.2  Humidity = 13.0%
Temp = 75.2  Humidity = 13.0%
Temp = 75.2  Humidity = 13.0%
Temp = 75.2  Humidity = 13.0%
Temp = 75.2  Humidity = 13.0%
Temp = 75.2  Humidity = 13.0%
Temp = 75.2  Humidity = 13.0%
Temp = 75.2  Humidity = 13.0%
Temp = 75.2  Humidity = 13.0%
Temp = 75.2  Humidity = 13.0%
Temp = 75.2  Humidity = 13.0%
Temp = 75.2  Humidity = 13.0%
Temp = 75.2  Humidity = 13.0%
Temp = 75.2  Humidity = 13.0%
Temp = 75.2  Humidity = 13.0%
```

Figure 9-3   AdafruitDHT.py script screen display.

Figure 9-4   Analog Devices TMP36 temperature sensor.

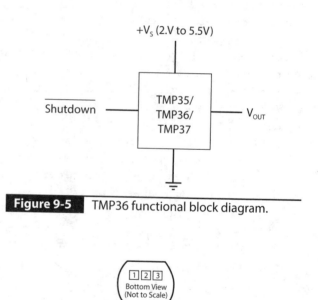

Figure 9-5   TMP36 functional block diagram.

1 2 3
Bottom View
(Not to Scale)

Pin 1   Pin 2   Pin 3
Vcc   Signal Out   Ground

Figure 9-6   TMP36 bottom view showing external leads.

far more complex than a simple transistor in that in contains circuits to both sense ambient temperature and convert that temperature to an analog voltage. The functional block diagram is shown in Figure 9-5. The TMP36 has only three leads, which are shown in the bottom view in Figure 9-6.

Table 9-1 provides details concerning these three leads, including important limitations.

Table 9-1 TMP36 Pin Details

Pin Number	Description	Remarks
1	$+V_S$	Supply voltage; ranges from 2.7 to 5.5 V
2	V_{OUT}	The analog voltage representing the temperature; the maximum voltage depends on the supply voltage
3	GND	Common reference used by both the supply and V_{OUT} pins

The voltage representing the temperature depends on the TMP36 supply voltage, which must be considered when converting the VOUT voltage to the equivalent real-world temperature. I do account for this in the software that converts the VOUT voltage to an actual temperature. Figure 9-7 is a graph of the signal pin voltage versus temperature using a 3-V supply voltage.

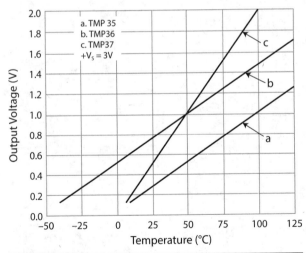

Figure 9-7 Graph of V_{OUT} voltage versus temperature for a $+V_S = 3$ V.

The actual temperature measurement range for the TMP36 is –40 to +125°C, with a typically accuracy of ±2°C and a 0.5°C linearity. These are good specifications considering that the cost of the TMP36 is typically less than $2. The TMP36 range, accuracy, and linearity are well suited for a home temperature monitoring system.

Analog-to-Digital Conversion

The RasPi does not contain any means by which analog signals can be processed. This means that an analog-to-digital converter (ADC) must be used before the RasPi can handle this sensor's signal.

I used a Microchip MCP3008, which is described on the Microchip datasheet as a 10-bit SAR ADC with SPI data output. Translated, this means that the MCP3008 uses a successive approximation register (SAR) technique to create a 10-bit digital result, which, in turn, is output in a serial data stream using the Serial Peripheral Interface (SPI) protocol. How the SPI protocol functions will be addressed after the sidebar that follows. The inexpensive MCP3008 ADC chip has impressive specifications despite its very low cost. Figure 9-8 shows the package form and pin-out for this chip.

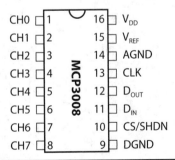

Figure 9-8 MCP3008 package form and pin-out.

The MCP3008 chip used in this chapter's project is in a dual-in-line package (DIP), which means that I had to use a solderless breadboard to interface it with the RasPi. I encourage you to read the following sidebar if you are interested in how the MCP3008 accomplishes the analog-to-digital conversion. There will be no loss of continuity if you choose to skip the sidebar, however.

Inner Workings of the MCP3008 ADC Microchip

I will refer to the MCP3008 functional block diagram shown in Figure 9-9 throughout this discussion. The analog signal is first selected from one of eight channels that may be connected to the input channel multiplexer. Using one channel at a time is called operating in a *single-ended mode*. The MCP3008 channels can be paired to operate in a *differential mode* if desired. A single configuration bit named *SGL/DIFF* selects single-ended or differential operating mode. Single-ended is the mode used in this project.

The selected channel is then routed to a sample-and-hold circuit, which is one input to a comparator. The other input to the comparator is from a digital-to-analog converter (DAC), which receives its input from a 10-bit SAR.

Basically, the SAR starts at a 0 count output and rapidly increments to a maximum of 1,023, which is the largest number that can be represented with 10 bits. Each count increment also increases the voltage appearing at the other comparator's input. The comparator will trigger when the DAC voltage precisely equals the sampled voltage, and this will stop the SAR from any further incrementing. The digital number that exists on the SAR at the moment the comparator triggers is the ADC value. This number is then output, 1 bit at a time, through the SPI circuit, which I discussed below. All this takes place between sample intervals. The actual voltage represented by the ADC value is a function of the reference voltage V_{ref} connected to the MCP3008. In our case, V_{ref} is 3.3 V; therefore, each bit represents 3.3/1,024, or approximately 3.223 mV. For example, an ADC value of 500 would represent an actual voltage of 1.612 V, which was computed by multiplying 0.003223 by 500.

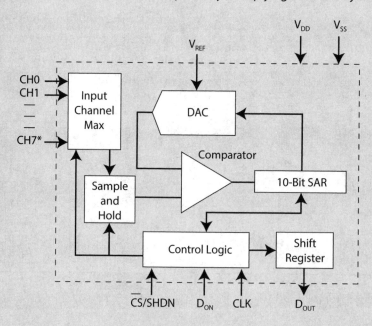

Note: Channels 4–7 are available on MCP3008 only.

Figure 9-9 MCP3008 functional block diagram.

Serial Peripheral Interface

The SPI is one of several data communication channels that the RasPi supports. It is a synchronous serial data link that uses one master device and one or more slave devices. A minimum of four data lines are used with SPI, and Table 9-2 shows the names associated with the master (RasPi) and the slave (MCP3008) devices.

Table 9-2   SPI Data Line Descriptions

Master Device (RasPi)	Slave Device (MCP3008)	Remarks
SCLK	CLK	Clock
MOSI	D_{in}	Master out, slave in
MISO	D_{out}	Master in, slave out
CS/SHDN	SS	Slave select

Figure 9-10 is a simplified block diagram showing the principal components used in an SPI data link. There are usually two shift registers involved in the data link, as shown in the figure. These registers may be hardware or software depending on the devices involved. The RasPi implements its shift register in software, whereas the MCP3008 has a hardware shift register. In either case, the two shift registers form what is known as an *interchip circular buffer arrangement* that is the heart of the SPI.

Data communication is initiated by the master by first selecting the required slave. The RasPi selects the MCP3008 by bringing the SS line to a low state or 0 VDC. During each clock cycle, the master sends a bit to the slave, which reads it from the MOSI line. Concurrently, the slave sends a bit to the master, which reads it from the MISO line. This operation is known as *full duplex communication*, that is, simultaneous reading and writing between master and slave.

The clock frequency used depends primarily on the slave's response speed. The MCP3008 can easily handle bit rates of up to 3.6 MHz if powered at 5 V. Because we are using 3.3 V, the maximum rate is a bit less at approximately 2 MHz. This is still very quick and will process the RasPi input without losing any data.

The first clock pulse received by the MCP3008 with its chip select (CS) held low and D_{in} high constitutes the start bit. The SGL/DIFF bit follows next and then 3 bits that represent the selected channel(s). After these 5 bits have been received, the MCP3008 will sample the analog voltage during the next clock cycle.

The MCP3008 then outputs what is known as a *low null bit*, which is disregarded by the RasPi. The following 10 bits, each sent on a clock cycle, are the ADC value with the most significant bit (MSB) sent first down to the least significant

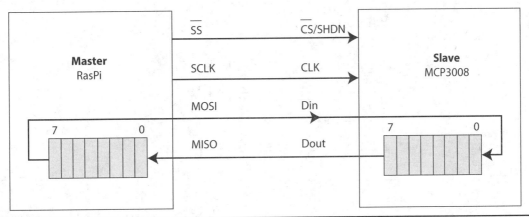

Figure 9-10   SPI simplified block diagram.

bit (LSB) sent last. The RasPi will then put the MCP3008 CS pin high, ending the ADC process.

Initial Test

Initial testing involves both creating a hardware circuit and establishing the proper Python software environment.

Hardware Setup. I will first discuss the hardware circuit because it is relatively straightforward. Figure 9-11 shows the test schematic for the T-Cobbler, MCP3008, and TMP36. I connected the TMP36 V_{out} lead to the MCP3008 Channel 0 input, which is pin 1, as shown in Figure 9-8.

The actual physical setup is shown in Figure 9-12. On the right side of the breadboard, you can see the TMP36 sensor connected with three jumper wires to the other breadboard circuitry.

Table 9-3 shows the equivalents between the ADC count, sensed temperature, and voltage. Use these values to verify that the TMP36 sensor is accurately measuring the ambient temperature. You can easily add a calibration factor if the measured temperature does not equal the

true temperature as measured by a separate calibrated thermometer.

Table 9-3   ADC Count, Voltage, and Temperature Equivalents

ADC value	Temperature (°C)	Voltage
0	−50	0.00
78	−25	0.25
155	0	0.50
233	25	0.75
310	50	1.00
465	100	1.50
775	200	2.50
1,023	280	3.30

Software Setup. You next need to load the Python developer libraries, which will allow you to support the script to run the SPI circuit. Install the Python development libraries by entering

```
sudo apt-get install python-dev
```

The following test script displays a continuous stream of temperature values generated by

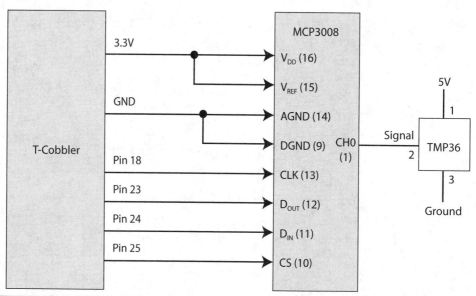

Figure 9-11   Test schematic.

Figure 9-12 Physical test setup.

the TMP36 sensor. The program is named *TMPSensor.py* and is available for download from this book's website, www.mhprofessional.com/NorrisHomeAutomation. The code follows the MCP3008 ADC configuration guidelines and SPI protocol, as discussed earlier. The code employs a "bit-banging" approach to SPI interface implementation. This approach makes running the script independent of any prerequisite SPI driver installations, which I felt made the test process as simple as possible.

```python
#!/usr/bin/env python

import time
import os
import sys
import RPi.GPIO as GPIO

GPIO.setmode(GPIO.BCM)

# read SPI data from MCP3008 chip, 8 possible channels (0 thru 7)
def readadc(adcnum, clockpin, mosipin, misopin, cspin):
    if ((adcnum > 7) or (adcnum < 0)):
        return -1
    GPIO.output(cspin, True)

    GPIO.output(clockpin, False)  # start clock low
    GPIO.output(cspin, False)     # bring CS low

    commandout = adcnum
    commandout |= 0x18  # start bit + single-ended bit
    commandout <<= 3    # we only need to send 5 bits here
```

```python
    for i in range(5):
        if (commandout & 0x80):
            GPIO.output(mosipin, True)
        else:
            GPIO.output(mosipin, False)
        commandout <<= 1
        GPIO.output(clockpin, True)
        GPIO.output(clockpin, False)

    adcout = 0
    # read in one empty bit, one null bit and 10 ADC bits
    for i in range(12):
        GPIO.output(clockpin, True)
        GPIO.output(clockpin, False)
        adcout <<= 1
        if (GPIO.input(misopin)):
            adcout |= 0x1

    GPIO.output(cspin, True)

    adcout >>= 1        # first bit is 'null' so drop it
    return adcout

# define a function to convert raw count to a voltage level
def ConvertVolts(data, places):
    volts = (data * 3.3) / 1023
    volts = round(volts, places)
    return volts

# define a function to calculate temp from TMP36 data
def ConvertTemp(data, places):
    temp = ((data * 330/1023) - 50)
    temp = round(temp, places)
    return temp

# change these as desired - they're the pins connected from the
# SPI port on the ADC to the T-Cobbler
SPICLK = 18
SPIMISO = 23
SPIMOSI = 24
SPICS = 25

# set up the SPI interface pins
GPIO.setup(SPIMOSI, GPIO.OUT)
GPIO.setup(SPIMISO, GPIO.IN)
GPIO.setup(SPICLK, GPIO.OUT)
GPIO.setup(SPICS, GPIO.OUT)

# TMP36 connected to adc #0
temp_adc = 0;
```

```
# define delay time
delay = 2

while True:

    # read the analog pin
    temp_level = readadc(temp_adc, SPICLK, SPIMOSI, SPIMISO, SPICS)
    if temp_level == -1:
        print 'incorrect ADC channel'
        sys.exit()
    temp_volts = ConvertVolts(temp_level, 2)
    temp = ConvertTemp(temp_level, 2)

    # display results
    print '-------------------------------------'
    print temp_level, '      ', temp_volts, '       ', temp

    # delay (in seconds) between measurements
```

Run the script by entering

```
sudo python TMPSensor.py
```

Figure 9-13 is screen shot of a portion of the program output with the TMP36 sensor measuring ambient room temperature. In the figure, the number in the left-hand column is the raw count coming from the MCP3008 ADC. The number in the middle column is the equivalent voltage for the raw count. The right-hand column shows the equivalent temperature for the voltage in degrees Celsius.

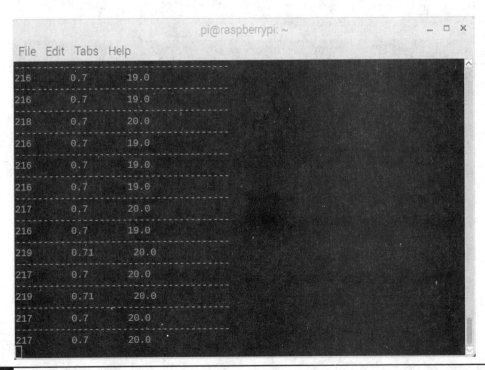

Figure 9-13 Initial test results.

Passive Infrared Sensor

This sensor is designed to sense the presence of humans or other warm-blooded mammals. Figure 9-14 shows a typical passive infrared (PIR) sensor, which is inexpensive and commonly used for both motion and occupancy detection.

Figure 9-14 Passive infrared sensor.

This sensor's ratings, pin connections, and range jumper settings are detailed in Figure 9-15, which comes from the manufacturer's datasheet.

The PIR sensor uses a crystalline material that generates an electric charge when exposed to infrared energy. The amount of energy generated is proportional to the size and thermal properties of nearby objects. Environmental conditions also affect the sensor, including ambient light and heat sources. A Fresnel lens located in front of the active element that focuses all incoming infrared signals. An onboard electronic amplifier is triggered when rapid infrared signal changes are detected. The detection range for this particular sensor may be modified by repositioning a jumper located in the upper left-hand corner, as shown in the quick-start circuit of Figure 9-15. One position

Pin Definitions and Ratings

Pin	Name	Type	Function
1	GND	G	Ground: 0V
2	Vcc	P	Supply Voltage: 3 to 6VDC
3	OUT	O	PIR Signaling: HIGH = Movement; LOW = No Movement

Pin Type: P = Power; G = Ground; I = Input; O = Output

Jumper Settings

Symbol	Description
S	Reduced sensitivity mode, for a shorter range, about 15 feet maximum
L	Normal operation, for a longer range, about 30 feet maximum

Quick-Start Circuit

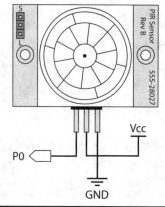

Figure 9-15 Sensor ratings, pin connections, and jumper settings.

is the so-called normal position and provides a nominal 30-foot detection range. The other position is the reduced-sensitivity one, where the detection range is half the normal range, or 15 feet. That position would be appropriate for indoor use in a normal-sized room.

This type of sensor is affected by ambient temperature, which should not be surprising considering that it basically works by detecting rapid temperature changes. Figure 9-16 is a graph that shows the effects of ambient temperature on the sensor for both the normal and reduced-sensitivity jumper settings.

Test Script

There are no special software dependencies that need to be installed in order to support this sensor. It functions just fine with the normal RPi.GPIO library that I have used in previous chapters, which is the only library needed to interface with the RasPi GPIO pins. The following is the Python script I wrote to test this sensor. I named this script *PIRTest.py*, and it is available from this book's companion website, www.mhprofessional.com/NorrisHomeAutomation.

```
#!/usr/bin/env python

import time
```

```
import RPi.GPIO as GPIO

GPIO.setmode(GPIO.BCM)

# connect to GPIO pin 18
PIR_pin = 18

GPIO.setup(PIR_pin, GPIO.IN)

while True:
    if GPIO.input(PIR_pin):
        print 'PIR alert'
    time.sleep(2)
```

Test Run

The sensor has some dim red LEDs in the Fresnel lens that will light for approximately 40 seconds when power is first applied to the sensor. This is the self-calibration period that the sensor requires for normal operations. It will be ready to run the script after this period ends and the LEDs go out. You can run the script by entering this command:

```
python PIRTest.py
```

Now wave your hand in front of the sensor, and you should see the message PIR alert appear in the terminal window. It may reappear because the sensor requires several seconds to reset after the initial motion ceases. The LEDs in the Fresnel lens will also light when motion is detected.

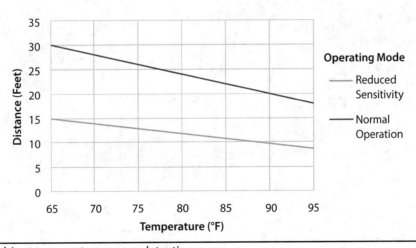

Figure 9-16 Ambient temperature versus detection range.

This test demonstrates that this type of sensor is quite simple in its operation and has little flexibility other than a range sensitivity setting. Nonetheless, this sensor type has been applied to a vast variety of home lighting applications, including driveway and porch lighting. However, the question naturally arises, what type of sensor is available if you need a more precise measure of a person's distance from the sensor? The next section addresses this issue.

Ultrasonic Sensor

An ultrasonic sensor provides for actual distance measurements between a target and the sensor. Figure 9-17 shows front and back views of the ultrasonic sensor used in this demonstration.

The ultrasonic sensor contains an embedded microprocessor as part of the encapsulated sensor hardware. This processor controls the ultrasonic transmitter and receiver transducers

Figure 9-17   Ultrasonic sensor front and back views.

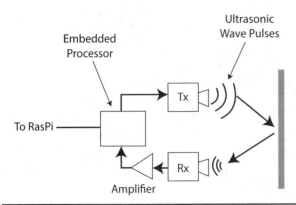

Figure 9-18 Ultrasonic sensor block diagram.

that physically measure distance by bouncing discrete ultrasonic sound wave pulses off objects and timing how long the sound pulse takes to make the round-trip transit. The distance is easily calculated because the speed of sound in air is relatively constant. This is very similar to how bats navigate in caves and attics. Figure 9-18 is a block diagram of the sensor showing how it functions.

The embedded processor generates an ultrasonic pulse when triggered by the RasPi. It also generates an output pulse if a return echo is detected. A RasPi GPIO pin is used to trigger a 40-kHz ultrasonic burst. The sensor then "listens" for an echo return and sets another RasPi GPIO pin high. Software running

on the RasPi detects the time differential between the trigger pulse and the echo return pulse and converts that time interval into an equivalent distance between the sensor and the reflecting target. The operational block diagram in Figure 9-19 illustrates this process and identifies the sensor and RasPi pins used for the interconnections. However, an important level-shifter chip is not shown in this figure but is shown in the schematic in the next section.

The ultrasonic sensor measures distances from 3 to 250 centimeters (cm) with an accuracy of approximately ±2 cm, which is less than a 1-inch error. Distance measurements also depend on the size and texture of the object that reflects the sound pulses. A wall provides excellent reflections, whereas a stuffed toy would be more problematic.

Physical Setup

This sensor requires 5 V for power and returns 5-V pulses for the echo signal. This voltage level is incompatible with a RasPi GPIO level input and must be reduced to 3.3 V or else damage will occur to the RasPi. I elected to use a level-shifter module to accomplish this reduction. Figure 9-20 is the interconnection schematic.

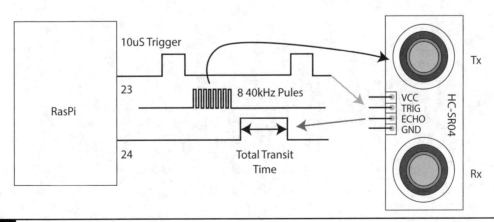

Figure 9-19 Operational block diagram.

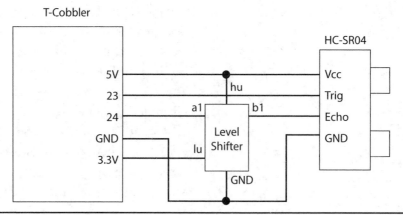

T-Cobbler

HC-SR04

5V → Vcc
23 → hu → Trig
24 → a1 | b1 → Echo
GND → Level Shifter → GND
3.3V → lu
GND

Figure 9-20 Interconnection schematic.

NOTE: You can also use a simple resistive divider to lower the input voltage to pin 24, but I already had a level-shifter module available and felt that it was a better solution to this problem. However, Figure 9-21 is a schematic for the resistive voltage divider for those of you who may choose to use that alternative.

Figure 9-22 shows the breadboard ready for a test. You should place the sensor near the outer edge of the breadboard, clear of any other components or wires. I placed a small box in front of the sensor as well as a ruler calibrated in centimeters between the sensor and the box target.

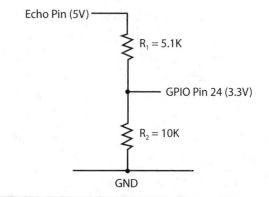

Echo Pin (5V)

$R_1 = 5.1K$

GPIO Pin 24 (3.3V)

$R_2 = 10K$

GND

Figure 9-21 Resistor voltage divider.

Figure 9-22 Physical setup.

Test Script

No special software dependencies are needed to support this sensor. It functions with the normal RPi.GPIO library, which that I have used in previous demonstrations. The following is the Python script I wrote to test this sensor. I named this script *UltrasonicTest.py*, and it is available on this book's website.

```python
#!/usr/bin/env python

import time
import RPi.GPIO as GPIO

GPIO.setmode(GPIO.BCM)

# define the GPIO pins
trigPin = 23
echoPin = 24

# setup the GPIO pins
GPIO.setup(trigPin, GPIO.OUT)
GPIO.setup(echoPin, GPIO.IN)

# set the trigger initially low
GPIO.output(trigPin, GPIO.LOW)

# short time to settle the sensor
time.sleep(1)

# forever loop
while True:
    # generate the 10uS trigger pulse
    GPIO.output(trigPin, GPIO.HIGH)
    time.sleep(0.000010)
    GPIO.output(trigPin, GPIO.LOW)

    # wait for an echo return
    while GPIO.input(echoPin) == 0:
        pulse_start = time.time()
    while GPIO.input(echoPin) == 1:
        pulse_end = time.time()

    # calculate pulse duration
    pulse_duration = pulse_end - \
        pulse_start
```

```python
    # calculate distance using a sound
    # velocity constant
    distance = pulse_duration * 17150

    # round-off distance to two dp's
    distance = round(distance, 2)

    # display the distance in cm
    print 'distance = ', distance

    # short pause between readings
    time.sleep(2)
```

Test Run

You can run the script by entering this command:

```
python ultrasonicTest.py
```

Figure 9-23 shows the script output using a good reflective target set at about 10 cm from the sensor. The displayed values very accurately correspond with the actual distance between the sensor and target. I then tried measuring the distance between the room ceiling and the sensor. I again found a very accurate measurement result.

The most popular form of object detection in HA applications has been with PIR sensors, which I discussed earlier. While reliable, they are easily activated and fooled by temperature or light changes, flying insects, and small animals. In contrast, ultrasonic detection is more reliable because it senses motion or presence of an object by the reflection of transmitted ultrasonic bursts. Ultrasonic presence detection can be a very useful tool in building automation and boosting energy savings through the control of lights, heating, and other energy consumers. Another potential HA application is parking detection, where a driver can be guided to very accurately park a car in a garage. Ultrasonic detectors also can be used in safety applications in which

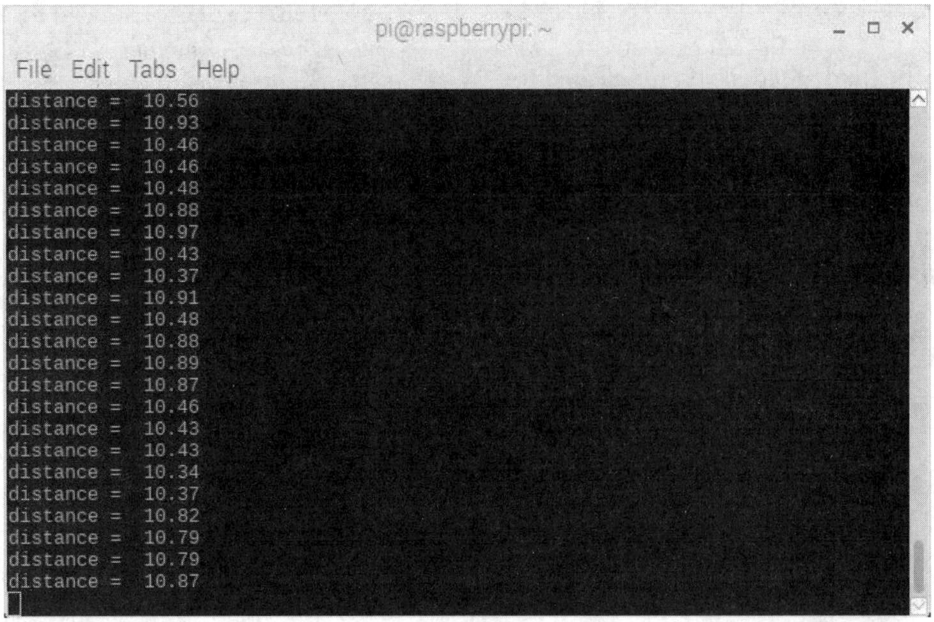

```
pi@raspberrypi: ~                              _  □  ×
File  Edit  Tabs  Help
distance =  10.56
distance =  10.93
distance =  10.46
distance =  10.46
distance =  10.48
distance =  10.88
distance =  10.97
distance =  10.43
distance =  10.37
distance =  10.91
distance =  10.48
distance =  10.88
distance =  10.89
distance =  10.87
distance =  10.46
distance =  10.43
distance =  10.43
distance =  10.34
distance =  10.37
distance =  10.82
distance =  10.79
distance =  10.79
distance =  10.87
```

Figure 9-23   Script output.

very reliable object detection is a high priority. A good application might be swimming pool surveillance, where it is critical to generate an alert or alarm if a small child attempts to enter an unattended pool.

Summary

This chapter discussed popular sensors often used in HA systems. My primary purpose was to explore how different sensors operate and how they can be interfaced with a RasPi.

The first sensor demonstrated was the DHT11, which is an integrated humidity and temperature sensor. It works using a software library provided by Adafruit, which has a single-line command to input a set of humidity and temperature readings from the sensor.

The next sensor described was a very inexpensive temperature sensor named TMP36. It outputs an analog DC voltage proportional to temperature. This situation requires the use of an analog-to-digital converter (ADC) between the sensor and the RasPi. I described how to set up a MCP3008 ADC to do the analog-to-digital conversion. I explained how the ADC used the SPI protocol to communicate with the RasPi. The project demonstration successfully measured ambient temperature using the TMP36, MCP3008, and RasPi.

I next demonstrated a passive infrared (PIR) sensor. This sensor uses a crystalline sensing element to detect changes in ambient heat signatures. It is very useful for motion-detection applications.

The last sensor demonstrated used ultrasonic acoustic pulses to measure distances between the sensor and a target. This technique is akin to how bats echo-locate around their environment. This sensor is very accurate and reliable and is independent of ambient temperature, which adversely affects PIR sensors.

## CHAPTER 10

HA Security Systems

SECURITY SYSTEMS ARE CURRENTLY a hot topic in the HA field. There are literally dozens of such systems on display at major home improvement stores and "big box" electronics outlets. I could not in good conscience write an HA book without discussing this topic. In order to properly discuss home security, I would first like to discuss the concept of risk because it has a big impact on the type of security system required for a specific situation.

Risk

Most readers have heard and used the word *risk* many times. Most probably have not thought too much about what it actually means and how it may be used in analyzing a particular condition. The online Merriam-Webster Dictionary definition of *risk* is

1: possibility of loss or injury: peril

Parts List

Item	Model	Quantity	Source
RasPi 3	B or B+	1	adafruit.com amazon.com mcmelectronics.com
Arduino development board	Uno Rev 3	1	adafruit.com amazon.com mcmelectronics.com
Google Home device	Any Home unit such as Basic or Mini	1	amazon.com
PIR sensor	555-28027	1	parallax.com
TauTronics Bluetooth transceiver	Version 4.1	1	amazon.com
XBee transceiver	XBee Pro S1	2	digikey.com mouser.com amazon.com
XBee Shield (Module)	—	2	sainsmart.com amazon.com
Bidirectional level-shifter chip	1875	1	adafruit.com

2: someone or something that creates or suggests a hazard

3a: the chance of loss or the perils to the subject matter of an insurance contract; *also*: the degree of probability of such loss

  b: a person or thing that is a specified hazard to an insurer

  c: an insurance hazard from a specified cause or source: war risk

4: the chance that an investment (such as a stock or commodity) will lose value

I prefer a simpler definition, which is actually contained in the same dictionary definition but is rephrased to highlight an important consequence:

Risk is the probability of a hazard happening.

This simple definition raises the question, what constitutes a hazard? Common sense dictates that a hazard is any unwanted event, which could be as minor as misplacing your car keys or as catastrophic as having your home burn down. Common security hazards likely encompass some of the following:

- Smoke/fire
- Water/flood damage
- Carbon monoxide (CO)
- Burglars
- Home invasion
- Loss of power
- Pet confinement
- Privacy

There are likely others that may be of concern, but let's stick with this short list. Each of the listed hazards typically has a probability associated with it. This probability is the chance that a hazard will happen. You will likely not know this value, which ranges from 0 to 1.0. A 0 value means that there is absolutely no chance

that a particular hazard will occur, whereas a 1.0 value means that it is 100 percent certain that a hazard will happen. For example, in the area where I live, it is common to have a loss of power occasionally during or after a severe storm passes through. Some probability values can only be determined by contacting appropriate local authorities. The local police chief or sheriff will likely have good statistics on the number of home burglaries and/or home invasions. Likewise, the local fire chief can provide good data on the number of home fires that have happened recently in your city or town.

Identifying potential hazards should always be the first step in creating an optimal security system tailored toward a specific home or business. The next step is collecting or otherwise assigning a probability to the identified hazards. Sometimes you must assign a value based solely on a "gut" feeling for the likelihood of a hazard happening. In any case, the next step is to whittle down the list and eliminate hazards that have a low probability of happening.

The next step in the process is to evaluate the cost versus benefit of mitigating the risk associated with all the hazards remaining on the list. Sometimes this is a rather easy task when considering risk versus cost. For example, mitigating the risk associated with smoke, fire, and CO detection in a home is just the purchase of a sufficient number of appropriate detectors and installing them in the recommended stations throughout the home. This risk mitigation is fairly inexpensive and, considering the hazard involved, an absolute necessity. In fact, most cities and towns in the United States at present require by government regulation that smoke detectors be installed in all new construction, whether it is residential or commercial.

Evaluating how much to spend on mitigating burglary risk is more problematic. It really depends on the crime prevalence in your community. If there have been many burglaries in your neighborhood, then the question becomes not if, but when. In this case, you should invest in a highly capable security system that will make your property less of a tempting target than other local properties. By contrast, if there have been no or extremely few burglaries in your area, then a minimal, inexpensive system likely would be appropriate. Notice that I did not say that no system was required because criminals always seem to be in ample supply and will eventually strike.

This discussion is simply a prologue to my first demonstration of a relatively simple security system that uses a single PIR sensor first described in Chapter 9.

PIR Security System

This system will implement a motion-detection system that can be placed high up in a room corner where it can cover the entire space. The remote PIR sensor is battery operated to facilitate easy placement and provide great flexibility to the user so that a room may be properly covered. I started this project with a list of requirements to clearly delineate what was

needed and to ensure that the system met the specific objectives required for implementing the room security.

Security System Requirements

The following is a list of the minimal requirements necessary to meet the goals of the security system:

- PIR motion detector
- Remote sensor, battery operated
- Wireless data transmission capability to a main RasPi controller
- Voice assistant activation
- Wireless alert/alarm transmission from the RasPi controller to the voice assistant

These requirements can be addressed in a variety of ways. I decided to break up the project into a series of subsystems to ease the overall design and construction.

Remote Sensor Hub

The components of the remote sensor assembly are shown in the block diagram in Figure 10-1.

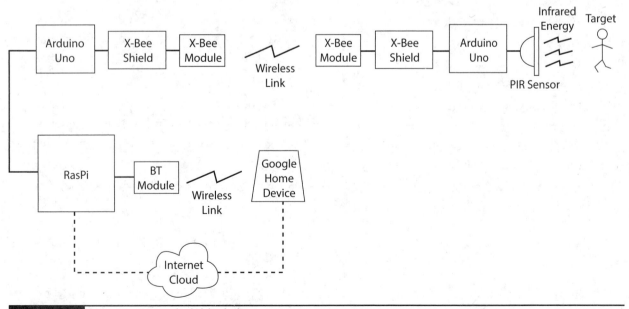

Figure 10-1 Remote sensor hub block diagram.

Note that I used two Arduino Uno Rev 3 boards as a controller for this the wireless data subsystem. An Uno controller is in direct wireless connection with the main RasPi controller using an XBee Shield, which, in turn, interfaces with an XBee RF module that implements the Zigbee data communications protocol. I discuss the XBee Shield, the XBee Module, and the Zigbee protocol in the next section.

The remote hub Uno controller will send an alert/alarm signal to the RasPi when it receives a signal from the PIR sensor on detecting motion within the room. The complete remote sensor assembly is power by a 6-V battery pack.

XBee and Zigbee Technologies

I selected XBee transceivers to implement the RF link because they are small, lightweight, inexpensive, and totally compatible with Uno boards. They also can transmit programmed digital data packets, unlike the RF transceivers discussed in Chapter 2, which only transmit several fixed and unchangeable digital data values. This additional flexibility is important in order to be able to transmit specific sensor data to support data-generator sensors. A data-generator sensor might include one that passes environmental data such as temperature and/or humidity.

XBee is the brand name for a series of digital RF transceivers manufactured by Digi International. Figure 10-2 shows one of the XBee Pro S1 transceivers that I used. There are two rows of 10 pins on each side of the module. These pins are spaced 2 mm apart, which is incompatible with the standard 0.1-inch spacing

Figure 10-2 XBee Pro S1 transceiver.

used on solderless breadboards. This means that a special connector socket must be used with the XBee Module to interconnect it with the Uno. This special socket is part of the XBee Arduino Shield, which is shown in Figure 10-3.

This shield can be purchased from sainsmart .com and contains all the functionality needed to effectively interface an Arduino board such as the Uno with an XBee Module. The shield and accompanying software make it very easy to create a useful RF communications link with very little effort.

The following sidebar examines the XBee hardware to show how this clever design makes wireless transmission so easy. Feel free to skip the sidebar if you are not particularly interested in learning about the XBee technology.

Figure 10-3 XBee Arduino Shield.

XBee Hardware

All the electronics in the XBee hardware, except for the antenna, are contained in a slim metal case located on the bottom of the module, as can be seen in Figure 10-4. If you look closely at the figure, you should see the bottom of the antenna wire, which is located near the top left corner of the case. While Digi International is not too forthcoming regarding what makes up the electronic contents of the case, I did determine that the earlier versions of the XBee Pro transceivers used the Freescale Model MC13192 RF transceiver. This chip is a hybrid type, meaning that it is made up of both analog and digital components. The analog components make up the RF transmit-and-receive circuits, whereas the digital components implement all the other chip functions. It is a complex chip, which is the reason why the XBee Module is so versatile and able to

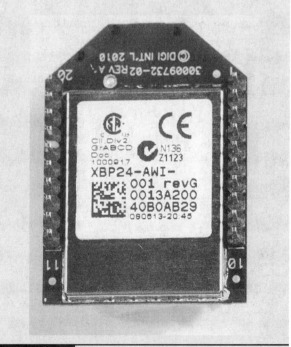

Figure 10-4 XBee electronics case.

(continued on next page)

automatically perform a remarkable number of networking functions. Table 10-1 shows a select number of features and specifications for the MC13192 chip.

Table 10-1 Freescale MC13192 Features and Specifications

Features/Specifications	Description
Frequency/modulation	O-QPSK data in 5.0-MHz channels and full spread-spectrum encode and decode (modified DSSS) Operates on one of 16 selectable channels in the 2.4-GHz ISM band
Maximum bandwidth	250 kilobits per second (kbps; compatible with the 802.15.4 standard)
Receiver sensitivity	≤92 decibel-milliwatt (dBm; typical) at 1.0 percent packet error rate
Maximum output power	0 dBm nominal, programmable from −27 to 4 dBm
Power supply	2.0 to 3.4 V
Power conservation modes	<1 μA of current 1 μA typical hibernate current 35 μA typical doze current (no CLKO)
Timers/comparators	Four internal timer comparators are available to supplement MCU resource
Clock outputs	Programmable frequency clock output (CLKO) for use by MCU
Number of GPIO pins	7
Internal oscillator	16 MHz with onboard trim capability
Operating temperature range	−40 to 85°C
Package size	QFN-32 Small Form Factor (SFF)

The XBee Module implements a full network protocol suite, but from a hardware perspective, this means that there must also be a microprocessor present in the electronics case. From my research, I cannot determine which type of microprocessor it is, but I am willing to make an educated guess that it is a Freescale chip based on the reasonable assumption that the MC13192 would be designed to be highly compatible with the company's own line of microprocessors. One other factor supporting my guess is that Digi International has recently introduced a line of programmable XBee Modules named *XBee Pro SB* that use the 8-bit Freescale S08 microprocessor.

The XBee pins are detailed in a logical arrangement in Figure 10-5 for your information. All the pin and function descriptions are shown in Table 10-2. Be aware that only four of the pins are needed for this project, and they are shown with an asterisk next to the pin number.

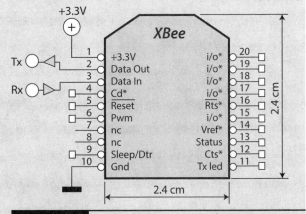

Figure 10-5 Logical XBee pin-out diagram.

Table 10-2 XBee Pin Descriptions and Functions

Pin Number	Name(s)	Description
1*	V_{cc}	Power supply, 3.3 V
2*	D_{out}	Data out (TXD)
3*	D_{in}	Data in (RXD)
4	DIO12	GPIO pin 12
5	Reset	XBee module reset, pin low
6	PWM0/RSSI/DIO10	Pulse-width modulation (PWM Analog 0), received signal strength indicator (RSSI), GPIO pin 10
7	DIO7	GPIO pin 7
8	Reserved	Do Not Connect (DNC)
9	DTR/SLEEP_RQ/DIO8	Data Terminal Ready (DTR), GPIO Sleep Assertion (pin low), GPIO pin 8
10*	GND	Ground or common
11	DIO4	GPIO pin 4
12	CTS/DIO7	Clear To Send (CTS), GPIO pin 7
13	ON/SLEEP	Pin high when *not* sleeping
14	Vref	Voltage reference level (used with analog-to-digital conversion)
15	ASSOC/DIO5	Pulse signal when connected to a network, GPIO pin 5
16	RTS/DIO6	Request To Send (RTS), GPIO pin 6
17	AD3/DIO3	Analog input 3, GPIO pin 3
18	AD2/DIO2	Analog input 2, GPIO pin 2
19	AD1/DIO1	Analog input 1, GPIO pin 1
20	AD0/DIO0/COMMIS	Analog input 0, GPIO pin 0, commissioning button

A considerable number of functions are available to you if needed, but this project requires only the most minimal functions for simple and reliable data transfers. Thankfully, the two XBee Modules automatically connect and establish reliable communications when power is applied to them. A red blinking LED on the XBee Shield is your indication that a communications link has been established.

I will finish this sidebar by mentioning that the XBee uses a highly capable networking protocol name *Zigbee*, which is also called a *personal area network* (PAN).

I fit the Uno board along with the XBee Shield and Module into a plastic case, as shown in Figure 10-6. The assembly is completely self-contained and is easily mounted to a wall surface using adhesive Velcro strips. I also decided to use a small solderless breadboard within the case to facilitate interconnecting all the components for the first prototype. It is always important to be able to easily reconfigure and reconnect components during initial testing to resolve any latent problems or issues. Figure 10-7 shows the electrical schematic for the complete hub assembly, which includes a PIR sensor.

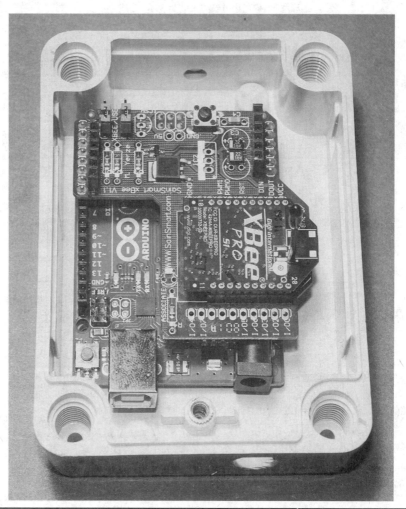

Figure 10-6 Remote sensor hub in a plastic case.

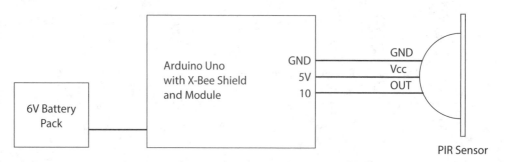

Figure 10-7 Remote sensor hub schematic.

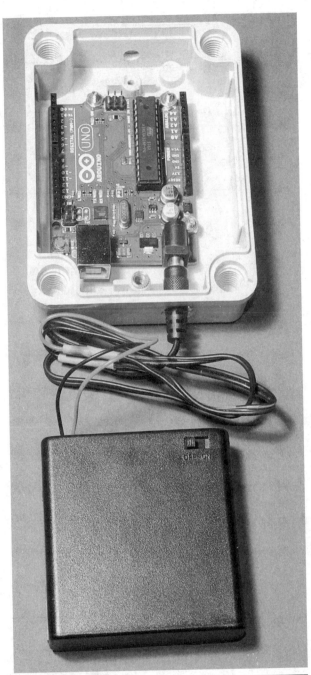

Figure 10-8 Battery pack connection to remote sensor hub assembly.

I also connected a coaxial plug to the battery pack, which inserts into the Uno board's coaxial power socket. I disconnect the power when not testing to conserve battery life. The Uno board with the Arduino Shield removed is shown in Figure 10-8.

I will discuss the remote sensor assembly software after the next section concerning the main RasPi controller.

Main Controller Assembly

I used a RasPi 3, Model B, as the main controller for this project. Figure 10-9 shows the block diagram for the main controller assembly.

The RasPi has an Uno board along with an XBee Shield and XBee Module, which provide two-way communication with the remote sensor hub. It also has an external Bluetooth transceiver that wirelessly connects to the voice assistant, enabling an audio stream to be sent from the RasPi to the voice assistant. The audio stream is sourced from the RasPi's 3.5-mm audio output jack. You may be wondering why I didn't just use the built-in Bluetooth connection hardware in the RasPi 3. The answer is that I did try it and found that the software installation was quite difficult and the actual wireless connection was just too unreliable. I decided that use of an external Bluetooth module was a much simpler and more reliable approach, and the module cost was also reasonable. Sometimes you just have to take an alternative route to proceed with a project.

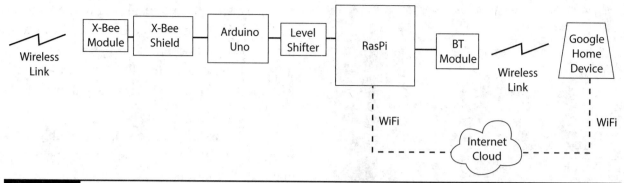

Figure 10-9   Main controller assembly block diagram.

The complete main controller assembly was set up on a tabletop for convenience without any attempt to package it as a complete assembly. It is powered from AC mains except for the external Bluetooth module, which has an internal rechargeable battery. However, I did plug the module's charging micro USB cable into one of the RasPi's USB sockets. The main controller schematic is shown in Figure 10-10.

I used a solderless breadboard with a T-Cobbler interface adapter to interconnect all the components for convenience and ease of assembly. I would have housed all the components in a plastic case if I had wanted to make this system a permanent item in the house. However, a tabletop version was adequate for this prototyping stage.

The remaining system component to be discussed is the voice assistant.

Voice Assistant

I elected to use the Google Home device to act as the voice assistant for this project. I decided on this unit because I liked the idea of using Web services with the RasPi main controller so that I could easily expand the system capabilities without having to fuss with the Alexa Skills, if I

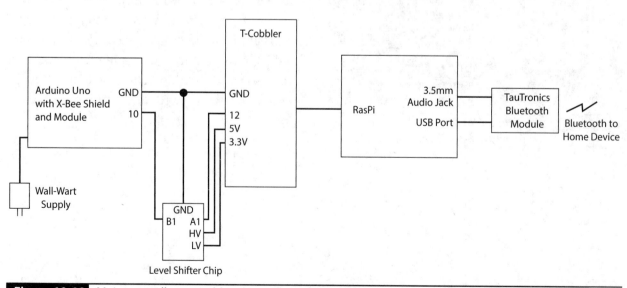

Figure 10-10   Main controller assembly schematic.

elected to go that route. Feel free to use an Alexa device if you so choose, but you will need to create your own Skill software to communicate with the Alexa device. That should not be too hard, provided that you follow the guidelines I presented in Chapter 4.

The Home device applet software is detailed within the following software generation and installation sections.

Software Generation and Installation

I have chosen to parse the software sections into separate groups, where each group addresses a major system component. The first component I discuss is the RasPi's main controller because that is the cornerstone of the complete system.

Main Controller Web Software

Creating the RasPi's Web software is a fairly simple process as long as you keep in mind what the subsystem requirements are and how they should be implemented. The requirements for the main controller are as follows:

1. Respond to a Web request to activate the system

2. Activate the sensor by sending a signal to the remote sensor assembly

3. Respond to any alert/alarm message sent by the remote sensor assembly

4. Send a present alert/alarm to the voice assistant via Bluetooth

I will comment on each one of these requirements as I go through the Web software development details.

In the spirit of rapid software development, I decided to reuse the Web server code I created in Chapter 2. I simply extended the existing code by creating a new snippet that would respond to any applet requesting that the PIR sensor be enabled. Please review the Chapter 2 discussion regarding applets and corresponding Web services if what I just mentioned does not make sense to you.

The newly extended Web server code listing, which I named *securityTest.py*, is as follows:

```
# Using RPi.GPIO and Flask for this script
import RPi.GPIO as GPIO
from flask import Flask

app = Flask(__name__)

# This is the default method that is invoked without an extension
@app.route("/", methods=['GET', 'POST'])
def index():
    GPIO.setmode(GPIO.BCM)
    GPIO.setup(12, GPIO.OUT)
    print "Turning the LED on"
    GPIO.output(12, GPIO.HIGH)
    return "LED on"

# This method is invoked when an "/off" extension is detected
@app.route("/off", methods=['GET','POST'])
def off():
    GPIO.setmode(GPIO.BCM)
```

```python
    GPIO.setup(12, GPIO.OUT)
    print "Turning off the LED"
    GPIO.output(12, GPIO.LOW)
    return "LED off"

# This method is invoked when an "/AClampon" extension is detected
@app.route("/AClampon", methods=['GET','POST'])
def AClampon():
    GPIO.setmode(GPIO.BCM)
    GPIO.setup(7, GPIO.OUT)
    print "Turning on the AC lamp"
    GPIO.output(7, GPIO.HIGH)
    return "AC lamp on"

# This method is invoked when an "/AClampoff" extension is detected
@app.route("/AClampoff", methods=['GET','POST'])
def AClampoff():
    GPIO.setmode(GPIO.BCM)
    GPIO.setup(7, GPIO.OUT)
    print "Turning off the AC lamp"
    GPIO.output(7, GPIO.LOW)
    return "AC lamp off"

# This method is invoked when an "/turnonsecurity" extension is detected
@app.route("/turnonsecurity", methods=['GET','POST'])
def EnableSecuritySystem():
    GPIO.setmode(GPIO.BCM)
    GPIO.setup(18, GPIO.OUT)
    print "Enabling the security system"
    GPIO.output(18, GPIO.HIGH)
    return "Security system enabled"

def DisableSecuritySystem():
    GPIO.setmode(GPIO.BCM)
    GPIO.setup(18, GPIO.OUT)
    print "Disabling the security system"
    GPIO.output(18, GPIO.LOW)
    return "Security system disabled"

if __name__ == "__main__":
    app.run(host='0.0.0.0', port=80, debug=True)
```

But wait, if you examined the listing, you realized it only controls GPIO pin 18 and has nothing in it that would activate the remote sensor hub. I did this on purpose because I have not yet shown you anything regarding how I implemented RF communications software and what type of data are linked between the main controller and the remote sensor hub. It is almost always a serious mistake to attempt to implement every requirement at one time. In this case, I will only light a LED connected to pin 18 as proof that the Web request is working as expected. The RF software portion will come later as I complete all the code.

The next logical step in code development is to create two applets that will request that the security system be enabled or disabled. Again, I will not repeat the clear instructions provided in Chapter 2 but simply summarize some of the information required to create the applets. One new applet is required to activate the security system and another one to disable it. I used the following example URLs for these applets:

```
http://mytestsite.org/turnonsecurity
http://mytestsite.org/turnoffsecurity
```

Please note these URLs are just examples, not real ones, so they will not work. The applets both use a POST method, which is how the desired action is directed to the appropriate method within the RasPi Web server.

The activation trigger phrases I used in the applets are

- "Turn on security system"
- "Turn off security system"

The two response phrases are

- "The security system is on"
- "The security system is off"

The main controller software is ready for a quick initial test prior to proceeding with adding more complexity to the system.

Initial Test   I connected a LED to GPIO pin 18 in preparation to test the newly extended RasPi Web server. I then started the RasPi Web server with this command:

```
sudo python securityTest.py
```

I then spoke the next phrase to test the "Turn on security system" applet:

- "OK Google, turn on security system."

The LED connected to pin 18 turned on after I spoke the phrase into the Home device. I also heard the phrase "The security system is on" come from the Home device, confirming that the Web request was completed. In a similar manner, the LED turned off after I spoke the phrase "OK Google, turn off the security system." As above, I heard the phrase "The security system is off" come from the Home device, confirming that the desired Web request was completed.

These actions confirmed that the Web server and applets were functioning as desired. It was now time to expand the software to include the RF communications link with the remote sensor hub.

RF Communications Link   My initial objective in implementing the wireless link was just to convey a state change whenever the PIR sensor detected motion. By far the easiest way to do this was by using GPIO pins on both the Uno controller and one RasPi GPIO pin. I was well aware that this approach runs counter to what I discussed earlier in having actual data sent, but my immediate goal was to quickly and reliably create a working RF link. The wireless link can be extended at a later time to handle real data exchanges, but for now, I just wanted to get the system up and running.

Remote Hub Hardware and Software Installation I already described the remote hub assembly in a previous section. I do have to provide a schematic showing how the PIR sensor connects to the Arduino board before describing the software that controls the whole assembly.

Remote Hub Hardware Installation Figure 10-7 showed the three-wire connection between the Uno and the PIR sensor. I mounted the PIR sensor to a small solderless breadboard, which was attached to the plastic case. The three interconnecting wires were routed through a hole on the top of the case to the Arduino board. Figure 10-11 shows the complete assembly with the PIR sensor in place.

You will be ready for the remote hub software installation once the preceding assembly steps have been completed.

Remote Hub Software Installation You will need to install the latest Arduino IDE in order

Figure 10-11 Complete remote hub with sensor.

to install the code described in this section. This is a free IDE download available for Windows, Mac, and Linux platforms at https://www.arduino.cc/en/Main/Software. I will not attempt to explain how to use this software because most of the readers of this book are quite likely to be very familiar with this IDE. However, a lot of information and tutorials are available on the Internet for readers who have never used an Arduino.

This next listing, which I named *pirTest.ino*, transmits either a 1 or a 0 to any other XBee in its vicinity. A 1 is sent whenever the PIR sensor detects an object in its field of vision; otherwise, a 0 is transmitted. The 1's or 0's are sent out once per second.

```
int inPin = 10;
byte value = 0;

void setup() {
  Serial.begin(9600);
  pinMode(inPin, INPUT);
}

void loop() {
  value = digitalRead(inPin);
  Serial.println(value);
  delay(1000);
}
```

This program is written using the Processing language, which for all practical purposes is a lightweight version of C/C++. The `Serial` class performs all the functions necessary to automatically interface the Uno with the XBee Shield. The statement

```
Serial.println(value);
```

is all that it takes to transmit the data values using the XBee Shield.

There is no need to go through a power-on sequence with the Uno board. Any uploaded program stored in its EEPROM will automatically run when power is applied to the Uno. Likewise, the Uno is designed to shut off "gracefully" when the power is disconnected.

No other software installations are required for the remote sensor hub. The next software installation to be discussed concerns the main controller assembly.

Additional Main Controller Hardware and Software Installations   I have already described most of the principal components required for the main controller assembly in a previous section. The Web software necessary to control the main controller assembly was also described and tested in previous sections. I also provided a schematic showing how the Uno with an XBee Shield attached connects to the RasPi. In addition, I described how to set up the external Bluetooth module, which creates a wireless link between the RasPi and the Home device.

Main Controller Hardware Installation Figure 10-10 showed how to connect the Uno to the RasPi. Please note that I used a level-shifter chip between the Uno and the RasPi because the Uno output level is 5 V, whereas the RasPi cannot handle any level greater than 3.3 V without likely damage. Both the 3.3- and 5-V supplies are sourced from the T-Cobbler adapter.

The Uno and XBee assembly are powered from a separate wall wart power supply. Do not attempt to power the Uno from the T-Cobbler 5-V power source. There is simply not enough current available from that source to power both the RasPi and Uno boards.

A TauTronic external Bluetooth module is shown in Figure 10-12. The 3.5-mm phone jacket shown in the figure must be plugged into the matching socket mounted on the RasPi.

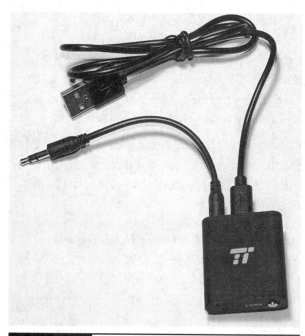

All the components were situated on a tabletop for convenience. Figure 10-13 shows the main controller assembly with all the components interconnected with a solderless breadboard.

You should now be ready for the main controller software installation once the preceding assembly steps have been completed.

Main Controller Software Installation   There are two scripts to be discussed regarding the main controller software installation. The first is the Arduino code that is stored in the UNO, which has an XBee Shield connected to it. The second portion concerns the Python code stored on the RasPi, which responds to any alert/alarm sent to it by the UNO. The RasPi will then send an alert via the Bluetooth module to the Home device.

Figure 10-12   TauTronic external Bluetooth module.

Figure 10-13   Complete main controller assembly.

I downloaded and installed the Arduino IDE on the RasPi from https://www.arduino.cc/en/Main/Software. Just ensure that you select the Linux Arm version from the list shown on the right-hand side of the website, as shown in Figure 10-14.

Using the Arduino IDE directly with the RasPi makes the overall process of creating and loading software into the Uno very easy. I likewise found making any Arduino software modifications a snap by having the IDE natively installed. I highly recommend that you use this approach versus developing the software on a separate PC and then connecting the Uno to the RasPi and testing it out.

The Uno software script, which I named *mainController.ino*, is as follows:

```
int outPin = 10;

void setup() {
  Serial.begin(9600);
}

void loop() {
  if(Serial.available()) {
    char c = Serial.read();
    digitalWrite(outPin, LOW);
```

```
    if(c == 85) {
      digitalWrite(outPin, HIGH);
    }
    delay(1000);
  }
}
```

The script is very simple, with the Uno constantly receiving characters sent from the remote sensor hub with the single statement

```
char c = Serial.read();
```

The `Serial` class once again makes XBee data links exceedingly easy, as I mentioned in the remote sensor hub software installation discussion.

The script tests for a trigger value, which when detected will set an Uno GPIO pin high. This pin is also connected through a level-shifter chip to a RasPi GPIO pin. The RasPi constantly monitors or polls its GPIO pin to detect when it changes state from low to high. A Python script installed in the RasPi will then send an alert to the Home device via the external Bluetooth module, which is connected to the 3.5-mm audio output jack.

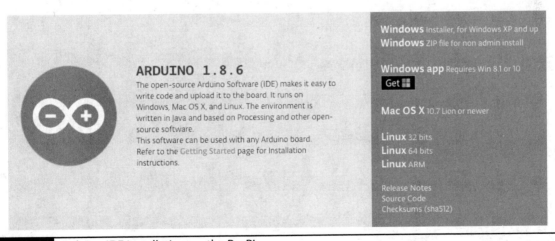

Figure 10-14 Arduino IDE installation on the RasPi.

The Python script, which I named *receiveTest.py*, must be loaded into the RasPi. This script is as follows:

```
import time
import RPi.GPIO as GPIO
import os

inPin = 12

GPIO.setmode(GPIO.BCM)
GPIO.setup(inPin,GPIO.IN)

while True:
    val = GPIO.input(inPin)
    if(val == 1)
        os.system('omxplayer -o local
                Wake.m4a')
    time.sleep(10.0)
```

This script is also quite simple. The designated GPIO pin is constantly polled to check whether its state has changed from low to high. If changed, a system-level call is made, which, in turn, sends an audio file to the Home device indicating that the PIR sensor has been triggered. I actually used a music file to send as an alert to the Home device. You can use any mp3 or m4a file you want as an alert.

Systems Test   I first tested the XBee wireless link by powering on the remote transceiver and monitoring the output using the integrated serial monitor. Just to reiterate, recall that the `Serial.println(value);` statement in the Arduino script causes the value both to be transmitted and also to appear on the serial monitor output screen, if so enabled. Figure 10-15 is a screen capture of the serial monitor screen.

You should be able to see the few 1's that appear on the screen when I waved my hand in front of the PIR sensor.

The next step in the systems test was to confirm that the main controller XBee transceiver was receiving and processing the data sent from the remote sensor hub. I again

Figure 10-15   Remote sensor hub serial monitor screen.

used the serial monitor to review the received characters. Figure 10-16 is a screen capture of that serial monitor being run on the host RasPi.

You should be able to see all the GPIO trigger events that I generated by waving my hand in front of the PIR sensor. The displayed messages result from debug print statements that I placed in the development code and are not shown in the book listing.

At this point I have confirmed a working wireless communications link between the remote sensor hub and the main controller. The next step is test the response action of the RasPi on detecting an event sent by the remote sensor hub. This would mean installing and running the Python script described earlier. However, you must first pair the TauTronics module with the Home device in order to determine whether the audio file is successfully transmitted from the RasPi to the Home device. I accomplished this pairing operation quite easily using the Google Home app installed on my smartphone. The process is quite simple in that you first put the Home device in pairing mode using the app. The app quickly discovers the TauTronics device, and you click on the app, confirming that you want to pair it. That's all that is needed for the pairing.

The Python script can now be run once the pairing operation is finished. Just load the script into the RasPi and run this command:

```
sudo python receiveTest.py
```

I was most pleased when I heard the alert "music" coming from the Home device after waving my hand in front of the PIR sensor. This last action completes the system tests, but I am not quite finished with this project because I have to discuss how to integrate the Web portion with the non-Web software.

Integrating All the Software

I provided a detailed discussion and demonstration of how to enable a security system using a voice command through a Home device. I next showed how to create a distributed security system in which a remote sensor triggered by infrared energy in a space could subsequently trigger an audio alert in a Home device. What I haven't discussed is how to tie the Web actions to the distributed security system. This is actually a difficult problem because of the nature of the two concurrently running processes. One process is the RasPi Web server, which is constantly monitoring an incoming HTTP port and responding to any valid Web

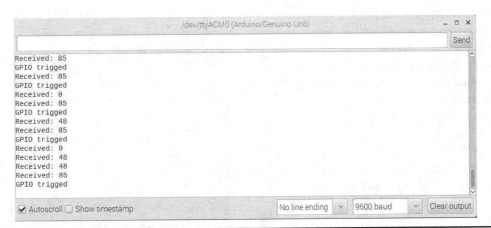

Figure 10-16 Main controller serial monitor screen.

requests sent by the Home device. The other process is essentially a forever loop in the Python script that is constantly polling to check on the state of a specific GPIO pin. These processes are mutually exclusive and basically cannot run simultaneously. After much experimentation, I finally arrived at a partial solution, which is to use a semaphore or flag, which would be used as a signal between the two processes. This flag is only a single character, either a 0 or a 1, and is stored in an external data file I named *outdata.txt*. The Web server would store a 1 in the data file whenever it received a Web request to enable the security system. Likewise, it would store a 0 in the data file whenever the disable Web request was received.

Meanwhile, the Python script was modified to read the flag from the data file and only take action if the flag was set to 1. The critical modification I needed was to execute the Python script from within the Web server software whenever the security system was enabled by voice command. This modification worked, and I was able to run the complete system and generate alerts whenever the PIR sensor triggered. Unfortunately, I was not able to reinstate the Web server process once I initiated the Python script process. This is likely due to some complex interactions taking place at the Linux kernel level. This limitation is not too critical because all you need to do is reboot the RasPi, and it will restart the Web server without any issues.

The modified Web server script is listed next. I renamed it *securityTestExtended* to differentiate it from the previous version. Note that I removed many of the URL extensions not required for this project.

```
# Using RPi.GPIO and Flask for this script
import RPi.GPIO as GPIO
from flask import Flask
import os

GPIO.setwarnings(False)

app = Flask(__name__)

# This method is invoked when an "/turnonsecurity" extension is detected
@app.route("/turnonsecurity", methods=['GET','POST'])
def EnableSecuritySystem():
    outf = open('outdata.txt', 'w')
    GPIO.setmode(GPIO.BCM)
    GPIO.setup(18, GPIO.OUT)
    print "Enabling the security system"
    GPIO.output(18, GPIO.HIGH)
    outf.write('{}'.format(1))
    outf.close()
    # Execute the Python script running the security system
    os.system('sudo python receiveTestExtended.py')
    return "Security system enabled"

# This method is invoked when an "/turnoffsecurity" extension is detected
@app.route("/turnoffsecurity", methods=['GET', 'POST'])
```

```
def DisableSecuritySystem():
    outf = open('outdata.txt', 'w')
    GPIO.setmode(GPIO.BCM)
    GPIO.setup(18, GPIO.OUT)
    print "Disabling the security system"
    GPIO.output(18, GPIO.LOW)
    outf.write('{}'.format(0))
    outf.close()
    return "Security system disabled"

if __name__ == "__main__":
    app.run(host='0.0.0.0', port=80, debug=True)
```

I have also listed the modified Python script below. It was renamed *receiveTestExtended.py* to differentiate from the earlier version.

```
import time
import RPi.GPIO as GPIO
import os

# Semaphore
outf = open('outdata.txt', 'r')
val0 = outf.read()

inPin = 12

GPIO.setmode(GPIO.BCM)
GPIO.setup(inPin,GPIO.IN)

while True:
    # Check if the GPIO pin is high
    val1 = GPIO.input(inPin)
    if val0 == '1' and val1 == 1:
        # Play the alert "music" through
        # the Bluetooth module
        os.system('omxplayer -o local
                Wake.m4a')
    time.sleep(2)
```

It is my belief that it would be possible to have both processes running concurrently by incorporating a multithreaded approach. This approach would be quite complicated and probably unwarranted considering that this is a maker-style project and should be kept to a moderate level of complexity.

Extensions and Modifications

This security system obviously can be extended by adding additional sensors. The Web side of the code would not have to change, but the Python script controlling the system would have to be extended to incorporate all new sensors. These extensions would be easy if the sensors were binary, meaning on or off, just like the PIR sensor. However, the XBee system is fully capable of sending significant data over the link instead of simple 1's and 0's. In this case, it would not be hard to use analog sensors with the security system, including temperature or humidity sensors. In addition, you could easily add distance-measurement sensors such as Lidar units, which I have discussed in some of my other project books.

The system software could also be modified to incorporate a logging feature, where system status would be stored periodically in a log file along with a time stamp. This would provide a useful long-term feature for users wanting to examine the system environment over a prolonged period.

Summary

I started this chapter on HA security systems with a discussion of the basic concepts to be considered when designing and building a home security system. The nature of risk was examined, and I pointed out that while some things must be protected, others can be left alone because any adverse event associated with those items would be of low probability and not cost that much.

A discussion regarding the chapter's project was next. This project involved a remotely mounted PIR sensor that was wirelessly connected to a RasPi controller. The whole system was turned on or enabled by a voice command spoken to a Google Home device.

I next discussed the overall system design, and the requirements were carefully detailed. Then a plan was created to meet those requirements. I also included a detailed discussion of the XBee electronics used to implement the wireless link

between the remote sensor hub and the RasPi main controller assembly.

Several sections followed that included discussions concerning implementation of the hardware and software required for the system. There was also a discussion of how to set up the RasPi Web server software and associated Web applets needed for the voice control aspect of the project. This part of the discussion was based largely on the material presented in Chapter 2.

I next went through a comprehensive procedure regarding the Arduino and RasPi software required to operate the system. An initial systems test was shown, and the system was first enabled with a voice command, the PIR sensor was triggered, and finally, an alert (actually a music file) was heard from the Home device. I finished the chapter with a discussion of possible system extensions and modifications that reasonably could be made to further enhance the system.

## CHAPTER 11

Integrated Home Automation Systems

I HAVE DISCUSSED A VARIETY of HA systems in this book. They hopefully will be useful on their own, but they are specific to certain tasks such home HVAC or security. This brief chapter should provide you with an overview of useful concepts needed to tie different systems together.

The term *integration* is often used to describe this tying operation. It is also common in the HA community or industry to consider specific HA system as subsystems when considering building an overarching HA control system. The question obviously arises as to how this integration can be achieved considering that various subsystems can have widely different controller architectures and underlying software platforms.

The answer is that there is no one solution that solves this problem. HA manufacturers have developed and marketed many different solutions, but most have tended to have the unsatisfactory approach of simply buying all the subsystems from their own brand and not having to worry about different protocols or hardware interfaces. I personally find this approach rather unpleasant and perhaps very expensive if you have already invested in a nonhomogeneous mix of subsystems. I propose a significantly different approach that I have borrowed from the field of software design patterns.

Adapters

We have all encounter adapters of one sort or another in our daily lives. One common adapter that you might not think as an adapter is the closed-captioning (CC) feature found on most modern TVs. CC is needed for TV users who are either hard of hearing or deaf. It adapts the normal TV audio output to a form suitable for the hearing-impaired user. Now consider the shapes shown in Figure 11-1.

Trying to fit these two shapes together is simply not possible. However, if you introduce a third shape into the figure, as shown in Figure 11-2, it suddenly is possible to fit all the parts together.

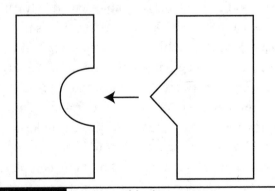

Figure 11-1   Incompatible shapes.

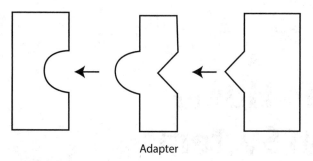

Adapter

Figure 11-2 Incompatible shapes with an adapter.

This new part is called an *adapter*, and it fixes the previous incompatible fit problem. The adapter solution is obvious when shown as a graphic in a figure, but it is not so obvious when considered as a programming or coding effort. The takeaway from this example is that the adapter has two different interface views depending on which side you view. Translated to programming terms, this means that the adapter code first receives incoming data in a specified format compatible with the data source. It then transforms that code into a format suitable for the outgoing data sink, inserting whatever "missing" pieces are required to match the output requirements. In most cases, the adapter code is bidirectional; that is, the output sink becomes the input source and the input source translates to an output sink.

There can be one big issue with this adapter approach, and it is directly related to how the communication links are implemented between data sources and sinks. There generally is no issue if data are sent and received using WiFi because the underlying data formats are totally compatible between all sources and sinks. However, if the data are communicated using a wireless specialized format such as analog RF, Z-Wave, or the XBee technology, then the adapter approach becomes a bit more complex. The data-source part of the adapter must then use compatible hardware to receive or transmit data to the desired subsystem. It isn't hard to do,

but it does raise the overall costs and difficulty in implementing the system. Fortunately, many commercial subsystems being manufactured today exclusively use WiFi for data links, so this is not a real concern. It is just something you should keep in mind when selecting subsystems for a total HA solution.

One-Stop Control

It only makes logical sense to have a single point of control for an HA system, or what I refer to as *one-stop control*. This control point can be implemented in a variety of ways from a tablet-like touch panel device as shown in Figure 11-3 to personal voice assistants such as the Amazon Echo device or the Google Home device.

Creating a one-stop solution has become quite easy if you elect to use a voice assistant such as the Home device. Figure 11-4 shows just a few of the dozens of websites that are available to support a vast array of manufacturer HA systems and services.

It has become an almost trivial effort to set up applets to voice control a vast array of diverse HA subsystems using the procedures I detailed in Chapter 2. I believe that a similar effort using touch panel control devices likely would be more difficult than voice-controlled devices. This is

Figure 11-3 Honeywell Tuxedo Touch HA system controller.

Figure 11-4 Sample of the dozens of readily available Web request websites.

due in large part to the immense popularity of the voice devices and HA manufacturers' desire to participate in this huge potential and actual market. There is one voice assistant device that combines both voice commands and video technology. This device is Amazon's Echo Show, which is pictured in Figure 11-5.

This device has ready access to dozens of manufacturer-provided Alexa Skills, which are basically equivalent to the applets created for the Google Home devices. Again, I believe that it would be very easy to create a one-stop solution using the Echo Show, which has an added advantage of including a visual component to HA system control.

Shared Sensors and Actuators

It makes perfect sense to share sensors among HA subsystems to help minimize cost and increase overall efficiency. For example, sharing light sensors between security and HVAC subsystems would likely be possible because both employ the same sensor types, but for different reasons. The success in sharing depends primarily on how a particular sensor connects with a given subsystem. For instance, if a sensor uses WiFi to broadcast or publish its

Figure 11-5 Amazon's Echo Show.

data, then any properly equipped subsystem can receive or subscribe to those data without issue or conflict with other subsystems. By contrast, sharing would be very problematic if a sensor is hardwired and uses a proprietary communications protocol to link to a particular subsystem. I also suggest that the chances of successfully sharing sensors would be greatly improved by selecting HA subsystems made by the same manufacturer. There always exist the possibilities of inherent incompatibilities in sensor usage even with wireless devices using standardized communications protocols such as WiFi or Bluetooth. HA manufacturers sometimes do not carefully comply with implementing published standards despite advertising that their sensors are compliant with a particular standard.

Sharing actuators is often an easier task than sharing sensors. This is due to the fact that most actuators simply respond to commands rather than generating data. For example, an architectural lighting control subsystem might employ a variety of light sources including incandescent, compact fluorescent, halogens, and LEDs, which may also be required to be dimmed and/or brightened to preset levels. A security subsystem likewise might employ the same types of light sources to illuminate an area based on any trigger events detected by the subsystem. Sharing the same lighting actuators between the systems is easily accomplished using well-known control circuits that provide isolation between connected subsystems but still accomplishing similar goals such as energy savings, providing visual interest, enhancing security, or just plain setting a mood for certain occasions.

Script Automation

It is often the case that an HA system provides a prescribed set of actions, which may be based on the time of day or user-requested actions. For instance, a user arriving home after a busy workday may very well want and deserve the HA system to set the home temperature to an appropriate occupancy level, start playing a favorite song through an AV system, close automated blinds/shades, set room lighting, and other user-desired tasks. The integrated HA system should follow a script that initiates these various subsystem tasks based on time of day, a trigger event, or manual activation. This type of script is also known as a *macro* in HA terminology and simply consists of a list of prestored subsystem commands with appropriate argument values. The commands are typically executed in serial fashion because some tasks are predicated on the successful completion of a prior task.

A script must be carefully constructed to accommodate abnormal events that can and will likely occur. For instance, it may happen that the AV subsystem has a problem preventing the favorite song from being played. This situation should not hinder the script from starting and completing all the other tasks contained in the script. The user should also be notified of the partial failure happening within the automated script.

One ideal goal of an automated script would be that a single button press or voice command could create the perfect ambiance for a dinner party, a romantic evening at home, or a party of friends on a back patio or deck. You should have no problem finding the appropriate music or video that you will enjoy when your AV equipment is managed by a HA subsystem. On command, the room lights dim, the shades close, and the appropriate equipment turns on. All you need to do is relax and enjoy the scene.

Additional Subsystems That May Be Automated

The previously mentioned subsystems are among the most popular ones to be controlled by an integrated and automated HA system. However, just about any product or system that uses electrical or battery power can also be integrated into an HA system. Some of these systems include but are not limited to

- Garage doors
- Swimming pools and spa systems
- Motorized security gates
- Video doorbells
- Electronic door locks
- Any motorized equipment (e.g., drapes, blinds, home theater screens)
- Irrigation systems
- Artificial ponds with waterfall features
- Decorative fountains

AI and HA

I already discussed an important AI topic, namely fuzzy logic (FL), in Chapter 8, where FL was used to control an HA HVAC system. However, AI can have a far greater impact regarding HA systems. Classification, prediction, and pattern recognition are some of the key functionalities that can be accomplished using AI. I will start with pattern recognition and how it might be used in an HA system.

Suppose that you repeatedly invoke a particular script every workday but also modify it by including some additional tasks such as playing a new song or perhaps controlling some extra lighting not included in the original script. An HA controller equipped with AI capabilities will notice the repeated modifications to an existing script and query you on whether or not you would like to modify the script such that you will not have to manually input the additional tasks each time the script executes. Pattern recognition is clearly involved in this process, and technically, another AI feature called *genetic programming* (GP) also would be involved. GP happens when a program automatically creates or modifies another program. Often GP uses random mutations in an effort to improve on an existing program in some fashion; however, in this situation, the modifications are already clearly defined and can take place without the need for random permutations.

Classification is another important AI topic that is also directly relevant to an HA application. Suppose that you have a video doorbell connected to an HA controller equipped for video image recognition. The system could then analyze a facial image and immediately invoke a script associated with that image. Imagine a situation where a couple shares a home. The husband or first significant other could arrive home first, and a script associated with that person would be run. Similarly, the second person in the relationship could arrive first, and a script associated with that person would be run. Of course, it would be a tossup if both arrived simultaneously, and the system might be programmed to do a random draw in order to determine which script would be run.

Video classification also may be appropriate in a security subsystem in which a sensor has been triggered by some unknown "intruder." The system could attempt to classify the intruder as human or nonhuman and take appropriate actions. If the intruder is human, there could be another attempt at facial recognition in order to ascertain whether the intruder was in fact a known entity or completely unknown. If unknown, the system would next take appropriate steps at notifying authorities and generating alarm sounds and/or voice alerts. A simple greeting would be generated if the target was identified. For nonhuman intruders, a warning sound might be generated to scare off the animal, unless the system included image recognition of family pets or obviously nonthreatening animals such as rabbits.

Prediction is the last AI topic I will discuss regarding an HA system. *Prediction*, as the name implies, means looking toward the future and trying to determine what is going to happen in the near term. A reasonable example of using prediction in an HA system would be employing an outside temperature sensor to help predict how to optimally operate an HVAC system. If the outside temperature is falling rapidly, the system could accurately predict that it might need to go into an aggressive heating mode. Similarly, if the temperature is rising rapidly, the system would counteract by entering a rapid cooling mode.

These examples just scratch the surface on how applied AI could improve an HA system. I recommend my book, *Beginning Artificial Intelligence with the Raspberry Pi*, if you want to learn more about AI and how it can be done using only a RasPi.

Index